Life in the Cold

LIFE IN THE COLD

An Introduction to Winter Ecology

PETER J. MARCHAND

with illustrations by Libby Walker

THIRD EDITION

University Press of New England
HANOVER AND LONDON

University Press of New England, Hanover, NH 03755

Printed in the United States of America 5

Library of Congress Cataloging-in-Publication Data

Marchand, Peter J.
Life in the cold : an introduction to winter ecology / Peter J. Marchand; with illustrations by Libby Walker. — 3rd ed.
p. cm.
Includes bibliographical references (p.) and index.
ISBN 0–87451–785–0 (alk. paper)
1. Cold adaptation. 2. Winter. 3. Ecology. I. Title.
QH543.2.M37 1996
574.5′42—dc20 96–19460
♾

TO THE LATE GARRETT CLOUGH

who pioneered this lifelong
pursuit with me,
never dreaming it would
come this far

TO GREG AND DANIELLE

who have, no doubt, gone without
for their father's obsession with science,
never quite understanding what it is
that I do for a living.

AND TO MY STUDENTS

of winter ecology past and future,
who always ask the right questions
and whose enthusiasm keeps me going
on the really cold days.

Take winter as you find him—and he turns

Out to be a thoroughly honest fellow with

No nonsense in him

And tolerating none in you

Which is a great comfort in the long run . . .

From Tass's winter cabin
James Russell Lowell

CONTENTS

PREFACE

It has always struck me as odd that traditional biology and ecology programs give so little attention to the special problems of plants and animals wintering in the North, especially considering that so much of our landscape lies under snow and ice cover for so much of the year. I suppose that there are many reasons for this. Traditional academic calendars, for one thing, do not allow much time to be spent in the field during the winter months. Even when time is available, the logistical problems of getting people and sampling equipment to study sites and of keeping that equipment (and sometimes the people) operating in the snow at subfreezing temperatures, are often rather discouraging. To me, however, this has been more a justification for the effort than a deterrent. And it's not that winter biology is unknown. To the contrary, there are a great many noteworthy research efforts recorded in the literature. The problem, though, is that the individual pieces are rarely integrated to make up the whole; and that, of course, is what ecology as a discipline is all about. So I have taken a small step here toward that end by piecing together a more comprehensive picture of organisms, both plant and animal, interacting with their environment and with each other in ways that are often unique to the season. The feelings of humble enlightenment as more and more pieces come together have made my winters much more enjoyable. This insight is the best that I can hope to pass on through the writing of *Life in the Cold.*

As in any discourse of a scientific or philosophical nature, it seems appropriate at the outset to define the bounds of the subject, if only to focus initial arguments. "Winter," to one degree or another, comes to every land beyond the tropics as the earth on its tilted axis alternately dips the Northern and Southern hemispheres toward and then away from the sun in making its annual rounds through space. In some places, the rhythm of life changes little

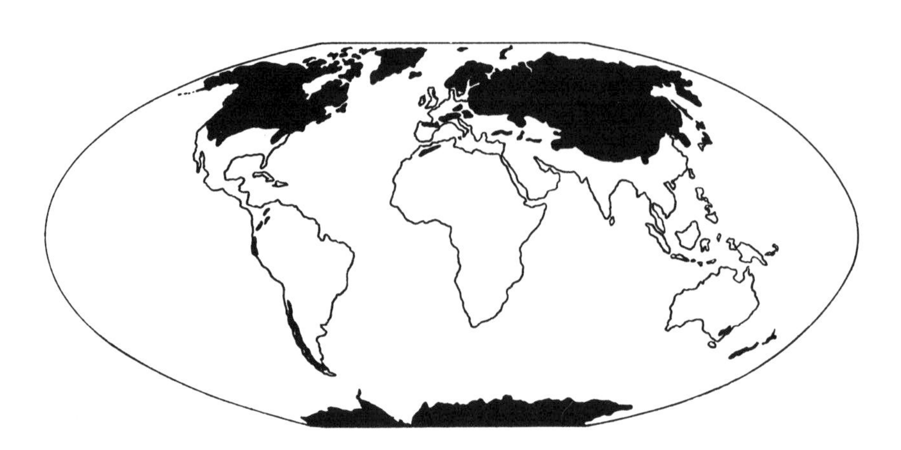

World snowcover distribution. Shaded areas are locations having snow on the ground for more than two months of the year, up to and including permanent snow and ice. (Data from G. T. Trewartha, An Introduction to Climate, *4th ed. [New York: McGraw-Hill, 1968].)*

through it all and one has to look carefully for the subtle signs that say this season is different from the rest. Elsewhere winter is the principal architect of life-form and habit. At higher elevations and higher latitudes, winter is a season of chilling energy deficits that demand the most conservative of physical and physiological adaptations in plants and animals. This is the winter I have in mind in writing this book. For the most part, then, I shall confine my discussions to those geographic areas where winter temperatures are low enough to elicit consistent and predictable *adaptive* responses from terrestrial organisms, and where snowcover is of sufficient duration to be considered an integral part of the environment (see map) and is important enough to make a difference in the overwintering success of the plant or animal. I shall let those two conditions define "winter" and prescribe the geographic coverage of this book.

This is by no means a conclusive treatise on the subject. Rather it is something of a progress report. I have been a student of winter ecology for nearly 20 years now, and I am just beginning to understand where the crucial matters lie. *Life in the Cold,* then, is as much a statement of what we don't know as what we do know. Throughout the book I have attempted, in a manner sometimes uncharacteristic of the sciences, to convey a sense of the freshness and personal excitement I feel toward the subject. In order to accomplish this, I have tried to reduce the formidable technicality of some material and remove, as much as possible without sacrificing accuracy, the veil of scientific jargon that sometimes inhibits, rather than helps, the learning process. In some cases I have stayed with familiar terms, even if they are less desirable from a scientific point of view, for example, where it seemed more convenient, retaining the terms "warm-blooded" and "cold-blooded" for their respective equivalents "homeotherm" and "poikilotherm." I have, however, used the international system of units throughout the text, hoping that it will not inconvenience my lay readers in countries still using English units. Despite my best attempts at simplification, it will be awkwardly clear in places that my primary commitment is still to the science rather than the prose.

Some may find *Life in the Cold* wanting in coverage of specific organisms. I have found it desirable, though, both in my writing and in my research, to limit my initial attention to problems of what I consider to be "winter-active" organisms—those that must face the rigors of winter on a day-to-day basis. I include plants as winter-active for reasons that will soon be obvious. In fact, early in my teaching I discovered that it is in the area of plants and the winter environment that the greatest misunderstanding (or lack of understanding) lies, and so I have treated certain aspects of plant ecology in considerable detail. On the other hand, because much more has been written elsewhere on the subject of animals in winter, I have confined my writing here to a few basic problems related mostly to the energetics of winter-active animals. If I succeed here in dispelling a few misconceptions and raising a few more unanswered questions that I think are important (or at least interesting), then I will be content for the moment. Once I have you thinking critically about winter ecology, then whatever deficiencies surface in my writing will no longer matter.

Though I take full responsibility for the content of *Life in the Cold,* I didn't get this far without a great deal of help. As a mentor and colleague of mine in the sciences often reminded me, "We all stand on the shoulders of the giants who walked before us." There have been many giants in my life. More than a few have lent a shoulder in the preparation of this book, some indirectly through their own contributions to the literature and others more directly through their enduring friendship and support, and through patient editorial and clerical assistance. Thanks to all, especially to Gerard Courtin of Laurentian University, whose critical review of this work went far beyond the call of duty. Thanks also to Richard Lee, Jr., Peter Pekins, and Robert MacArthur for valuable comments on materials added to later editions of this book. My debt is great.

Now it is my sincerest hope that the reading that lies ahead of you will enrich your own experience in the winter environment as much as the learning has enriched mine.

P.M.

UNIT CONVERSION FACTORS

To convert from	*to*	*multiply by*
kilometers	miles (statute)	.621
kilometers	feet	3280.84
meters	feet	3.281
meters	inches	39.37
centimeters	inches	2.54
hectares	acres	2.471
kilograms	pounds (avdp.)	2.205
grams	ounces (avdp.)	.035
liters	cubic feet	.035
liters	gallons (U.S. liquid)	.264
liters	quarts	1.057
milliliters	cubic inches	.061
milliliters	ounces (U.S. liquid)	.034
joules	calories (gm)	.239
joules	watt-seconds*	1
watt-hours	calories (gm)	860.421

*Note that the watt is a rate of work or power—specifically, that power that gives rise to the production of energy at a rate of 1 joule per second.

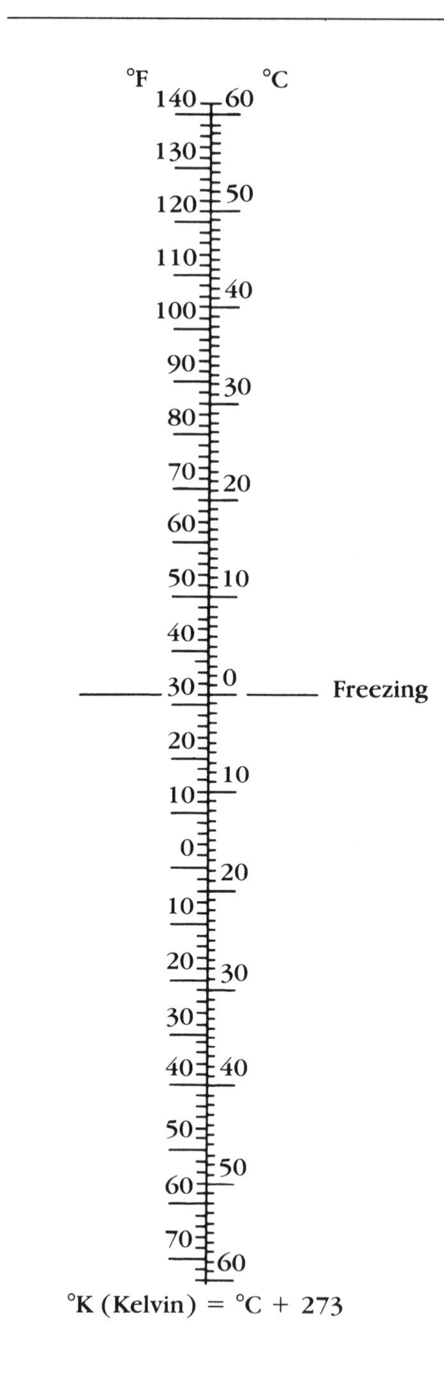

°K (Kelvin) = °C + 273

Life in the Cold

1 WINTER PATHS: OPTIONS FOR OVERWINTERING SUCCESS

In the lair under the uprooted old spruce, a heart beats faintly but steadily as the season marches on. All is quiet now, yet there is so much life in that still, dark hulk. Soon, barely waking, she will give birth to two or three cubs that will suckle in their mother's warmth until they are as full of energy as the sunlight bouncing off the late-winter snow. It is just one of many landmark events passing unnoticed in a natural world that has paused from its usual business to carry out a few of its more important acts in relative privacy.

Maybe the hardest notion to accept about winter is that it is so alive. Beneath the bark of the leafless tree, under the frozen moss, in all the little crevices of winter, there is life! For all the apparent calm of the winter landscape, a continuous play of activity is enacted daily behind the great white curtain. A masked shrew hastily nuzzles its way through the leaf litter, knocking down the fragile ice crystals that hang from the ceiling of the protective snowpack, almost frantically searching out food to fuel its metabolic furnace. Spiders, too, remain active under the snow, and with a little luck the shrew finds plenty to eat. But others have the same thing on their minds and there's nothing secure about the life of a shrew under the snowcover. With no more than a flick of a shadow to forewarn its prey, an ermine, diving down through the snow as adeptly as a cormorant at sea, comes up with the shrew and returns to line its own nest with the fur of its quarry (the ermine is warmed twice for its hunting skill).

The ermine, in its yellow-white winter pelt, is as natural to this scene as the snow itself. Almost from childhood we are taught that here is a creature truly at ease in the northern winter. But why? Its body is really all wrong for cold climates. Long and skinny, this animal is poorly adapted for conserving heat. While its short fur may keep ladies of fashion warm during their ever-so-brief encounters with winter, it hardly seems adequate insulation for subzero nights on the hunt. On the other hand the ermine *is* white, and we have the feeling that white in winter is good. But again, why? Protective coloration perhaps; but what does such a swift and voracious carnivore that spends much of its time in winter hunting the dark recesses under the snowpack need of protective coloration? Could it be that white confers some other not-so-obvious advantage?

As an object of winter ecology, perhaps we could choose no better symbol than the ermine, for it seems to represent all that is right *and* wrong in a winter-active animal. It reminds us that, within limits dictated by certain physical realities, there are many compromises in nature—that the forces shaping life in snow country often evoke different strategies, none of which solve all problems, but each of which serves a particular organism in terms of its own role in the scheme of things. Even among plants we find many alternative, often starkly contrasting solutions to the problems of winter. As we look closely, there too we find some common threads. Certain inventions work, others don't, and it is by trial and error that they are sorted out.

Through the endless process of natural selection, three basic strategies for surviving winter's rigors have evolved: migration, hibernation, and resistance. There are variations in theme, of course, but in one form or another these are the options. What they amount to in an evolutionary context is a choice between avoidance and confrontation. In a broader sense, though, each of these options confronts the season in its own way; even the "choice" of leaving a frozen landscape in favor of more southern latitudes is, in reality, a desperate stand against winter.

MIGRATION

At first glance, migration would appear a safe alternative to wintering in the North—an escape clause, as it were, in the contract between an organism and a demanding environment. But migration is a very precarious business. To begin with, the energetic cost of traveling long distances is extremely high. In order to fuel a 1000-km flight, a bird preparing for a migratory trip must accumulate so much reserve energy that it may carry at departure up to 50% of its total body weight as fat.[1] Adding weight for flight has its practical limitations, however, and even this preparation may not be enough.

The energy reserves of migrant birds flying over water sometimes become exhausted before they reach their destination.

Various other obstacles also confront the migrating bird. For some waterfowl there is the additional stress imposed by humans through unrelenting hunting along the migratory route. A Canada Goose leaving James Bay in early September, headed down the Atlantic or Mississippi Flyway, may face up to four months of shooting as the legal goose season opens progressively later southward along the route. When the survivors of this journey arrive at their destination, they face still other uncertainties. A new assemblage of parasites, diseases, and predators, along with changing food availability and difficulty in forming efficient new search habits, may take an additional toll.

Migration, thus, is not an easy out; but for many bird species there is no alternative. Physical or behavioral adaptations to particular feeding strategies alone may dictate fall flight. The herons, for example, with their stilt-legged manner of fishing for a living in shallow water, have no way of coping with even a thin, temporary cover of ice. They have, in effect, become too specialized. The flycatchers as well, once their insect prey have metamorphosed and become sedentary for the winter, must move southward to find food on the wing. And so, too, must the soaring birds of prey. When the northern landscape becomes uniformly white under snow and ice cover, there is no longer any differential heating of the earth's surface to generate uplifting air currents so that these predators may ride the thermals with minimal energy expenditure while they hunt.

Migration, then, is a necessary risk for some organisms. For others it is not a viable option at all, if only because the cost of transportation may be too great. In energetic terms, it is far more expensive to travel overland than to fly. It has been estimated that a mammal would expend approximately 10 times more energy moving a given distance by running than would a bird of equal weight flying that same distance.[2] This may be one reason why mammals in the North generally do not migrate to avoid winter. Interesting exceptions to this generalization include: caribou, which often move several hundred kilometers between their summer breeding grounds on the tundra and their winter shelter in the boreal forest; species of bats that migrate considerable distances to reach their winter refuge (using a more efficient means of getting there, of course); and such marine mammals as the gray whale, which may cover the entire Pacific coastline of North America at an energetic savings even greater than that of flying.

HIBERNATION

The alternative to migration as a means of avoiding the problems of food scarcity and extreme cold is hibernation, yet even this is a relatively uncommon overwintering strategy among mammals in the North. The true hibernator possesses an unusual ability as a "warm-blooded" animal or homeotherm (an organism that normally maintains body temperature independently of surrounding temperature) in that it can periodically enter a state of much reduced metabolic activity in which it allows body temperature to fall many degrees below normal, but without the debilitating effects usually attending hypothermia. Hibernation is an ability that has been perfected in only a few mammal species and in no birds.

While the advantages of this overwintering strategy will become apparent in later discussions, hibernation is not without its risks, as it requires animals to survive for an uncertain period of time on finite energy reserves. These reserves not only must be adequate for the maintenance of life functions such as respiration, but also must supply the energy for periodic arousals and rewarming during the winter. In addition, the safety of a hibernating animal might be limited in some areas by the availability of den sites in which the temperature is not likely to drop significantly below freezing. Only the arctic ground squirrel has been known to survive freezing in certain of its body tissues without sustaining permanent damage[3] (though some bats are suspected of having similar tolerance).

As in migration, the energy demands of hibernation may be quite high. No hibernating mammal maintains a continuously low

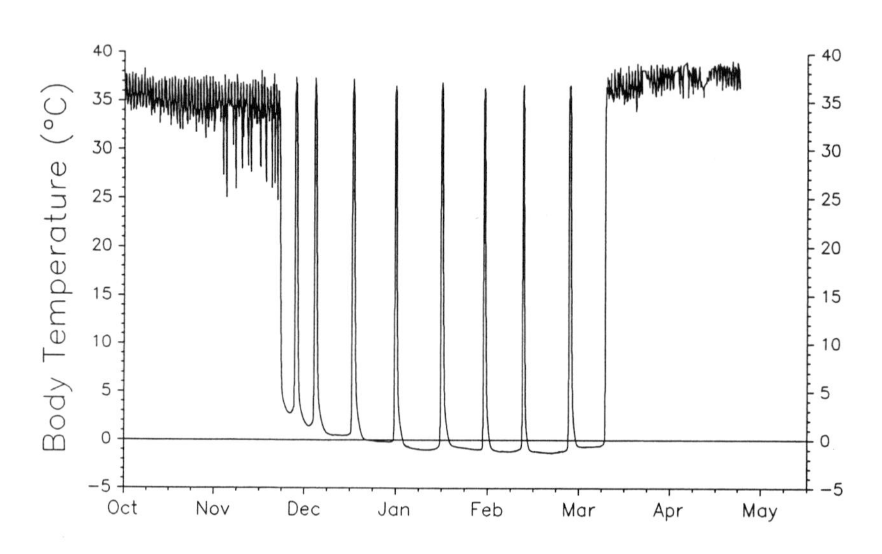

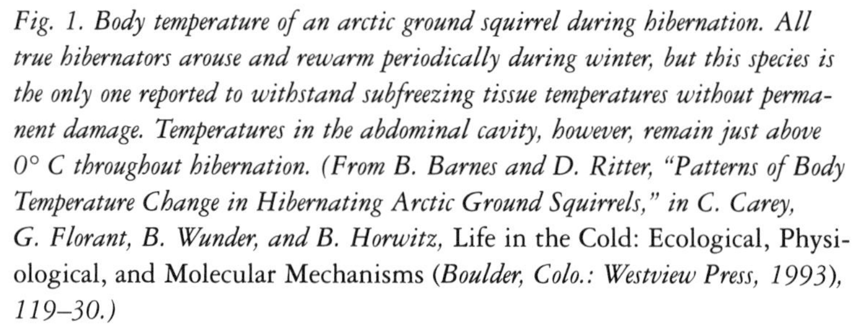

Fig. 1. Body temperature of an arctic ground squirrel during hibernation. All true hibernators arouse and rewarm periodically during winter, but this species is the only one reported to withstand subfreezing tissue temperatures without permanent damage. Temperatures in the abdominal cavity, however, remain just above 0° C throughout hibernation. (From B. Barnes and D. Ritter, "Patterns of Body Temperature Change in Hibernating Arctic Ground Squirrels," in C. Carey, G. Florant, B. Wunder, and B. Horwitz, Life in the Cold: Ecological, Physiological, and Molecular Mechanisms *(Boulder, Colo.: Westview Press, 1993), 119–30.)*

body temperature during winter, but rather, all arouse periodically, perhaps once every few days (fig. 1), even in the absence of freezing threat (there is growing consensus that the reason for arousal is to satisfy a sleep debt accumulated during hibernation, for sleep apparently requires normal body temperature).[4] The high energy cost of rewarming is subsidized by stored fat or stored food, representing a potential constraint on overwintering success should either prove insufficient for the term of hibernation. It is not surprising, therefore, that during particularly hard winters, mortality rates among small animals like jumping mice and ground squirrels may be quite high, ranging from 23 to 68% for adults and from 41 to 93% for juveniles.[5]

Hibernation clearly works for many animals, however, and our bear in its lair under the uprooted spruce tree represents perhaps the ultimate expression of wintertime escape. Though it maintains a body temperature only 6° to 7° C below normal (its large size and highly insulative fur preclude much more of a temperature drop), a denning bear reduces its metabolic rate by one-half and remains inactive for many months without eating, urinating or defecating. And their overwinter success rate may be better than 99%. For the bear, hibernation is as safe as life gets.[6]

For reptiles and amphibians of snow country the options are much more limited. These "cold-blooded" animals or "poikilotherms" have little control over their body temperature, and when the days and nights get cold their metabolism slows to the point of near helplessness. For the majority of these organisms, hibernation is the only means of surviving the winter season. As the days of fall become shorter, reptiles and amphibians seek out safe winter hiding places—sometimes under decaying logs, in deep protected rock crevices, or in the burrows of other animals—wherever they can escape subfreezing temperatures. Snakes sometimes congregate in groups of mixed species: rattlesnakes, copperheads, and northern black racers are frequently found together in rock dens. In unusual cases, countless thousands of a single species will congregate in deep caverns (such as those found in the limestone country of southern Manitoba), some individuals traveling 16 km to reach the same hibernaculum year after year.[7]

Frogs and toads may also seek underground burrows for hibernation, the latter frequently digging themselves into soft soil with their hind feet to a depth of nearly 1 m. Some have been known to use caves, too. Others take advantage of stream banks and seepage areas where temperatures are not likely to drop below freezing, or burrow into the soft mud of pond bottoms, sharing this winter refuge with salamanders and turtles. Occasionally reptiles and amphibians overwinter with little more protection than a cover of moss or leaf litter, and there they have been observed to survive temperatures of a few degrees below freezing. In fact, as we will soon see, some species of land frogs tolerate ice formation in the intercellular spaces of their body tissues, while protecting the living cells with glucose or glycerol (see chapter 4). But exposure to subfreezing conditions puts most in a precarious position. Any disturbance while in a "supercooled" state, where body tissues are dangerously chilled without ice formation below their actual freezing temperature, may result in immediate death. As was the case with hibernating mammals, the overwintering success of reptiles and amphibians is best assured by avoidance of subfreezing temperatures.

RESISTANCE

To many organisms winter means staying and enduring, facing the rigors of the season and resisting its stresses. Consider the sedentary tree or the leafless shrub with ice-covered buds poking above the snow. Are these totally dormant in winter, insensitive to the chill winds of January? Do they pass the season in frozen animation with their biological clocks set to resume life upon some proper cue in the spring? Or do they remain sensitive to their environment, able to respond to favorable circumstances during the winter? And consider the caterpillar overwintering in the frozen woodpile. Its emergence in the spring tells us it has remained very much alive. Yet these and so many other insects in their larval and pupal stages must withstand temperatures that would kill most living cells. Because these organisms produce no appreciable heat, they must depend upon complex biochemical mechanisms that enhance their ability to survive freezing temperatures. The production of glycerol by some insects, for example, inhibits ice formation in their body tissues, permitting them to cool without freezing, sometimes to temperatures as low as −50° C. In others the production of special ice-nucleating proteins promotes extracellular ice formation and reduces the risk of flash freezing, which is always a danger with supercooling. We think of these organisms as hibernating to avoid winter, but in reality theirs is a complex strategy for *resisting* severe cold stress (see chapter 4).

For winter-active birds and mammals, resistance often involves coping with snow and ice cover in the conduct of their daily affairs. Some animals are superbly adapted to the physical constraints imposed by snow, while others are substantially impaired by it. Lemmings develop a digging claw to aid them in foraging under the snow, and spruce grouse grow extra scales on their feet to enable a better grip while feeding on icy branches. The caribou, with feet seemingly way out of proportion to its body size, floats over the snow as if on snowshoes (fig. 2*a*). The lynx and snowshoe hare also benefit from the efficiencies of a high foot-surface-to-body-weight ratio. The moose, on the other hand, confronts deep snow differently, with a stature and musculature that allow it to lift its feet almost shoulder high, thus minimizing the amount of energy expended in wading through the snow. In marked contrast, the white-tailed deer suffers immeasurably in deep snow, unable to stay on the surface and unable to lift its legs clear (fig. 2*b*). The track of the white-tail shows characteristic foot drag in as little as 15 cm of snow. This inability to deal with deep snow explains in large part the yarding and feeding behavior of deer in northern areas during winter (see chapter 7).

In all, there exist a number of interesting adaptations in the animal world for coping with snow. But for many organisms the more serious problems of winter are associated instead with extreme cold, and for their survival a dependable annual snowcover is of utmost importance. Snow is both the reflective cover that turns away the sun's warming radiation and the thermal blanket under which much biological activity takes place during winter. While it is the bane of those plants that must bear its weight and those animals that must scratch through it for sustenance, it is the salvation of many other plants and animals that depend upon it for protection from the cold. Snow separates two worlds, ours and the "subnivean," that world beneath the snowcover. So important is it to both

a *b*

*Fig. 2. Coping with the deep snow. Caribou and their domesticated counterparts, the reindeer(*a*), are known as "floaters"—their oversized feet enable them to move easily over snow-covered ground. In contrast, the mobility of the white-tailed deer (*b*) is greatly hampered by deep snow, forcing them often to congregate in large "yards," usually under coniferous forest cover. Their collective trampling of the thinner snowcover here may allow more freedom of movement, but food resources soon become scarce in the vicinity of the yard and starvation is often their fate. (*Photo credits: 2a *by Peter Marchand;* 2b *by John Hall.)*

environments that William O. Pruitt, Jr., the well-known biologist from the University of Manitoba, considers the initial accumulation of 20 cm of snow to mark the true beginning of winter—the "hiemal threshold," as he terms it. It seems appropriate, then, that this present treatment of winter ecology should also begin with the subject of snow, considering first the physical characteristics of the snowpack and then probing the true nature of the environments above and beneath the snow surface.

2 THE CHANGING SNOWPACK

Snowflakes, for all their elaborate detail, are destined for destruction almost from the time they form. The molecules that make up the delicate needles and plates of multifaceted crystals are too active to remain long in such intricate formation. There are plenty of outside forces, too, that act upon the snowflake to hasten its destruction. Tumbled in the wind, fractured and compacted, vaporized and recondensed, and melted and frozen again, snow on the ground is constantly undergoing change or metamorphism. This not only alters the individual ice crystal, but also changes the internal structure of the snowpack.

Collectively, snow crystals are affected by at least three distinct processes in their metamorphism, and each process is influenced by time, internal snowpack characteristics, and external weather conditions. These processes are described here in the sequence in which they might occur after initial deposition, though it is possible for the three processes to occur in any order—or even simultaneously—within different parts of the snowpack.

DESTRUCTIVE METAMORPHISM

The initial deterioration of the snowflake and resultant formation of more or less rounded ice grains in the snowpack is referred to as "destructive" or "equi-temperature" metamorphism. The term "destructive" is self-explanatory as it pertains to the fate of individual snow crystals within the snowpack. The qualifier "equi-temperature" serves to distinguish this process from the next, which requires an unequal temperature distribution within the snowpack. Destructive metamorphism involves a reordering of water molecules on the surface of each snowflake with a resulting loss of fine structure—a deterioration of the radiating arms of delicate needlelike crystals, for example. This reorganization in effect represents a redistribution of molecular energy within the ice crystal. The snowflake is thus transformed into a roughly spherical particle with a net reduction in surface area. The resulting ice grains may then coalesce with others, the smaller ones being absorbed into larger ones, until all are nearly

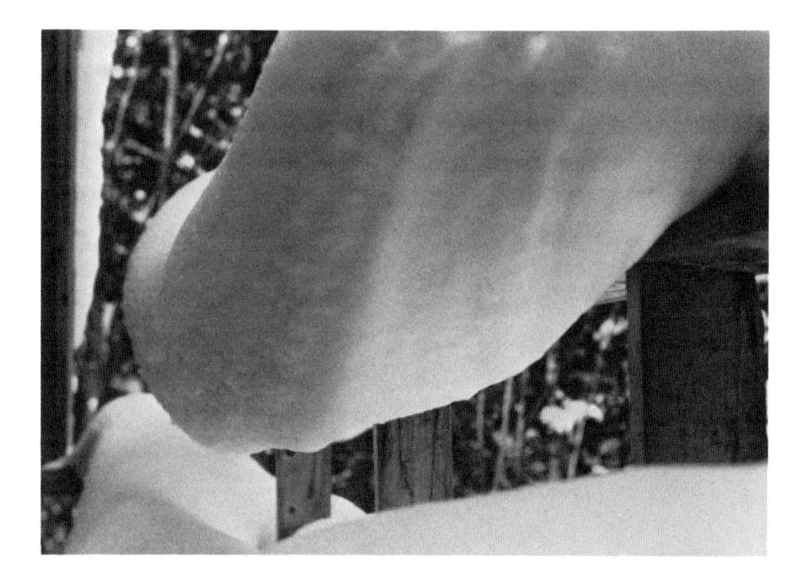

Fig. 3. Snow metamorphism. The increased mechanical strength of snow due to the bonding of individual ice grains—a process called "destructive metamorphism"—is evident in this picture of snow on a railing.

the same size. This process is influenced by anything that promotes closer contact of the ice crystals, such as wind packing or the weight of overlying snow, and is affected somewhat by temperature. Destructive metamorphism proceeds faster when air temperature is relatively warm than when air temperature is very low.

Throughout the process of destructive metamorphism snowpack air spaces are reduced in size as individual ice grains pack together and bond at their points of contact. Both the snowpack density and the mechanical strength of the snow increase substantially by this process, often within just a few hours after a snowfall (fig. 3). This is of some practical interest as it pertains to the use of snow shelters in the North. The traditional shelter of the Inuit winter camp, the igloo is constructed of wind-packed snow that owes its strength to the bonding that takes place during this metamorphic process. The effectiveness of the Quin-zhee, a snow house of Athapaskan origin used in boreal forest regions, is also related to the increase in strength accompanying destructive metamorphism when loose snow is piled and allowed a short while to "set" naturally (see box on "Snowpack Metamorphism and Throwaway Housing").

CONSTRUCTIVE METAMORPHISM

After the initial deposition and destructive metamorphism of snow on the ground, the snowpack begins to undergo changes in vertical structure through a process termed "constructive" or "temperature-gradient" metamorphism. Constructive metamorphism results from the migration of water vapor upward within the snowpack and depends on two snowpack properties: the existence of a temperature gradient between the bottom and top of the snowpack, and an interconnecting system of pore spaces within the snowpack.

A temperature gradient normally exists in the snowpack because of the low thermal conductivity of snow (a topic to be discussed in more detail later). The uppermost portion of the snowpack is affected by the cold air above, while the bottom of the snowpack is influenced by the ground, which provides a continuous source of heat throughout the winter. Because heat is conducted only very slowly through the snow, a pronounced temperature gradient is usually maintained between the colder top and warmer bottom of the snowpack. This temperature difference is important in that it affects the distribution of water vapor in the snowpack.

Within the pore spaces of the snowpack, water molecules are continually escaping from the ice surfaces until the internal atmosphere is crowded with vapor molecules and as many return to the ice surfaces by random bouncing as leave. Through this continual process of sublimation (the conversion of ice directly to water vapor) and condensation, the air spaces of the snowpack remain saturated—in other words, at 100% relative humidity. Because the amount of vapor in suspension at any given time increases with temperature, the warmer regions of the snowpack will have a higher concentration of water vapor molecules for a given volume of pore space than will

SNOWPACK METAMORPHISM AND THROWAWAY HOUSING

A Quin-zhee is a snow house of Athapaskan origin used in colder boreal forest regions of North America where snow is generally loose lying and of very low density. The shelter is constructed simply by shoveling snow into a large pile (*a*) and leaving it to set, without any additional packing, for an hour or two before hollowing out the inside living quarters (*b*). The

a

b

domed outside of the structure becomes self-supporting after a very short time because of a rapid increase in bonding strength between ice crystals, all due to the acceleration of destructive metamorphism caused by the mixing process and the weight of the snow itself. A person working alone can pile sufficient snow for a Quin-zhee in 30 minutes, and may begin digging out the interior only an hour or so later. A snowshoe is the only tool needed to construct a shelter that can warm to 25° above the outside air temperature. A very efficient means of obtaining shelter in deep fresh or unpacked snow, the Quin-zhee is easy to build and just as easy to leave behind, lessening the burdens of winter travel in the northern woods. (Photos by Peter Marchand.)

the colder portions. This means that, as long as there is a temperature gradient in the snowpack, there will also be a vapor concentration gradient. And because the molecules in suspension are continually bumping each other and trying to escape their confines, water vapor will tend to diffuse from areas of higher concentration to areas of lower concentration, which in this case means from warmer to colder regions of the snowpack. This migration via interconnected pores will go on even though the air spaces at the colder end of the gradient are saturated.

With the upward migration of water vapor through the snowpack, two things happen. First, condensation occurs in the upper, colder regions of the snowpack because the internal atmosphere is already at 100% relative humidity and the addition of more water vapor can only "supersaturate" the air. Condensation onto the outside of the ice grains in this region thus results in a growth of the grains. Secondly, water vapor migrating upward is continually replaced in the lower regions as sublimation from ice crystals maintains a saturated atmosphere there. This means that ice crystals at the bottom of the snowpack are continually diminishing in size.

Fig. 4. Depth hoar. As water vapor constantly migrates upward from warmer to colder regions of the snow, brittle, loosely arranged crystals known as "depth hoar" form at the base of the snowpack. Through this process of constructive metamorphism, more free space is created in the subnivean environment, facilitating the movement of small mammals, but at the same time contributing to avalanche danger in mountainous terrain. (Photo by Peter Marchand.)

The end result of constructive metamorphism is the formation, at the base of the snowpack, of very brittle, loosely arranged crystals referred to as "depth hoar" (fig. 4). The formation of depth hoar quite likely facilitates the movement of small mammals as they forage under the snow throughout the winter. Constructive metamorphism is also accompanied by a substantial loss of mechanical strength of the snowpack, and this has another consequence of considerable human concern. In mountainous regions the process of constructive metamorphism leads directly to an increase in avalanche danger. In times past when mountain troops were brought out to foot-pack threatening slopes, their job was, in effect, to reverse the constructive metamorphic process and promote destructive metamorphism instead.

MELT METAMORPHISM

The third process of snowpack alteration is "melt metamorphism," an almost self-explanatory change that comes about when any part of the snowpack is subjected to temperatures above freezing. As simple as it may appear, some of the finer points of this process are worth elaborating.

Surface melt may influence the internal temperature of the snowpack in a way that would seem out of proportion to the amount of melt taking place. This is due to the release of latent heat as meltwater percolating down into the snowpack encounters lower temperatures and refreezes. As already noted, under typical winter conditions the snowpack is warmest at the bottom and coldest at the top. However, as the air above the snowpack warms above freezing, the surface snow will become nearly the same temperature as the snowpack bottom, with a colder layer sandwiched between. This is related again to the fact that snow conducts heat relatively slowly, and changes occurring in the upper part of the snowpack may not affect lower portions right away. When surface meltwater percolates downward and refreezes in the colder stratum, it releases heat in the amount of 335 joules/g of water frozen—the latent heat of fusion (latent heat is the energy gained or lost with phase changes in water). The meltwater essentially acts as a heat pump, gaining 335 joules/g at the surface when it changes from solid to liquid and transferring this energy to lower layers of the snowpack when it freezes again. This has the effect of bringing the whole snowpack rapidly to equal temperature.

Melt metamorphism also occurs when rain falls on the snow surface or when fog forms over the snowpack. In the case of rainfall, the amount of heat energy transferred to the snowpack is related to the temperature of the rain itself and, of course, to the total quantity of rain entering the snowpack. These two measures give the total energy input of the rainwater, to which is added the latent heat released when this water freezes. In the case of fog, the amount of energy transferred to the snowpack is equal to the latent heat released as water condenses over the cold snow to form the fog droplet. The latent heat of condensation is quite high, amounting to approximately 2450 joules/g of water, depending on the exact air temperature. This means that, for every gram of water vapor condensed over the snow surface, enough heat energy is released to melt seven times as much ice (remember, the latent heat of fusion is only 335 joules/g). This helps explain why snow seems to disappear so much faster in fog than it does during a rain.

It should be apparent by now that the winter snowpack is a very dynamic entity, everchanging in its physical character. The processes described above continually alter the snowpack—constructive metamorphism breaking down old crusts, melt metamorphism forming new ones, fresh snowfalls starting the process over again with destructive metamorphism. As long as there is snow on the ground, these forces are at work. The result is a snowpack layered with a graphic history of winter's weather. Studying that record carefully may yield considerable insight into the problems experienced by organisms living on and beneath the snowpack.

THE INSULATIVE VALUE OF A SNOWCOVER

It has already been suggested that the winter survival of many organisms depends on the presence of an insulating cover of snow. The question arises, then, as to how much snow is enough. Is 20 cm, Pruitt's hiemal threshold, the critical depth? Or is more needed? The answer depends not upon depth alone, but also upon the degree of snowpack metamorphosis that has taken place. For among the various physical properties of snow that are altered by the processes of snowpack metamorphism the one that perhaps most affects life in the subnivean environment is changing snow density. Through its effects on thermal conductivity, density differences strongly influence the insulative value of a snowcover.

The thermal conductivity of snow is measured as the amount of heat passing through it over a period of time, for a given gradient in temperature between the top and bottom of the snowpack. The

Table 1. Thermal conductivities of various natural and man-made materials

Material	*Density* *(g/cm³)*	*Conductivity* *(W/cm/°K)*
Granite	2.70	.029
Ice (0° C)	.92	.021
Sandstone	2.25	.017
Concrete	2.85	.012
Glass	2.60	.008
Old snow	.40	.004
Wet peat	.90	.0033
Dry sand	1.55	.0021
Dry peat	.45	.0008
Sawdust	.19	.0004
Glass wool	.06	.0004
Fresh snow	.10	.0003
Still air	.001	.0002

Sources: R. Geiger, *The Climate Near the Ground* (Cambridge: Harvard University Press, 1965) and *Handbook of Chemistry and Physics* (Boca Raton, Fla.: CRC Press, Inc.).

lower the thermal conductivity of a material, the better its insulative value. A list of thermal conductivities for various natural and man-made materials is given in table 1. Notice that the thermal conductivity of snow is not only very low relative to many other natural materials, but it is also closely related to snowpack density. The higher the density of snow, the higher its thermal conductivity and, therefore, the poorer its insulative quality.

The insulating effect of a snowcover is most readily apparent in the reduction of daily temperature fluctuations underneath the snowpack. This effect can be assessed indirectly through calculation of a thermal index value that combines measurements of both depth and density (see box on "Estimating the Insulative Value of a Snowpack"). As thermal index values approach 200, the subnivean en-

ESTIMATING THE INSULATIVE VALUE OF A SNOWPACK

A quick and easy assessment of the insulative quality of a snowcover is possible through the use of a thermal index scale that integrates measurements of depth and density. Mathematically the index, I_T, is a simple summation of depth/density values for each layer of the snowpack, written as follows:

$$I_T = \sum_{i=1}^{n} (z/G)_i,$$

where z is the thickness (cm) and G is the density (g/cm^3) of each layer, i.* This expression simply says numerically what we have already seen about snow: its insulative value increases (hence, thermal index values will be higher) under snow layers of lower density or increasing thickness.

In practice, the thermal index is obtained by exposing a fresh vertical face of the snowpack, measuring the thickness of each discrete snow layer, and dividing the thickness by the density of the layer. The resultant values are then added for the whole profile. A snow sample may be collected with any kind of cylindrical corer (a small can opened at both ends works well) pushed either vertically or horizontally into the snowpack. The measured volume of snow thus obtained may then be weighed in a small plastic bag suspended on a spring balance to obtain snow density. With commercial coring devices such as the "Mt. Rose" or "Adirondack" snow samplers, the entire cylinder containing a vertical snow core is cradled on a balance, which is usually calibrated to convert snow density directly into water content. In the latter case, the thermal index can be approximated by sampling the snowpack with a single vertical core and dividing the total depth by a single density value. Thermal index values have proven useful in assessing the ecological significance of a snowcover, especially where it is marginal, or in early fall and late spring when snow conditions may be critically important to small mammals. It is a useful technique also for evaluating the impact of snow vehicles or similar disturbances on the quality of the subnivean environment.

*P. J. Marchand. "An Index for Evaluating the Temperature Stability of a Subnivean Environment," *Journal of Wildlife Management* 46 (1982): 518–20; and N. M. Kalliomaki, G. M. Courtin, and F. V. Clulow, "Thermal Index and Thermal Conductivity of Snow and Their Relationship to Winter Survival of the Meadow Vole, *Microtus pennsylvanicus*," *Proceedings, Eastern Snow Conference* 29 (1984): 153–64.

vironment is no longer influenced by short-term temperature changes above the snowpack (fig. 5). Further increases in thermal index value through the addition of more snow will make little difference to the temperature stability of the subnivean environment. It is interesting to note here that a thermal index value of 200 could result from a single snowfall of 20 cm depth where the density is .1 g/cm^3. This corresponds closely to Pruitt's hiemal threshold. Keep in mind, however, that the density of new fallen snow will increase over time through destructive metamorphism lessening its insulative value accordingly. On the other hand, once snow depth approaches 50 cm, changes in density up to .3 g/cm^3 become relatively unimportant in terms of their effect on subnivean temperature fluctuations because the greater depth compensates for the increased density. This is illustrated in figure 6, in which several snowpack temperature profiles, obtained with implanted thermistors, are plotted over a period of time during which considerable air temperature fluctuations occurred. Under 40 to 50 cm of snow, the temperature of the subnivean environment is almost constant.

For many small mammals, the presence of an adequate snowcover is critically important to their overwintering success. Restricted in their ability to add insulating fat and fur, they depend instead on the snowpack. Often their only means of maintaining a favorable heat balance in the face of decreasing air temperatures is to minimize

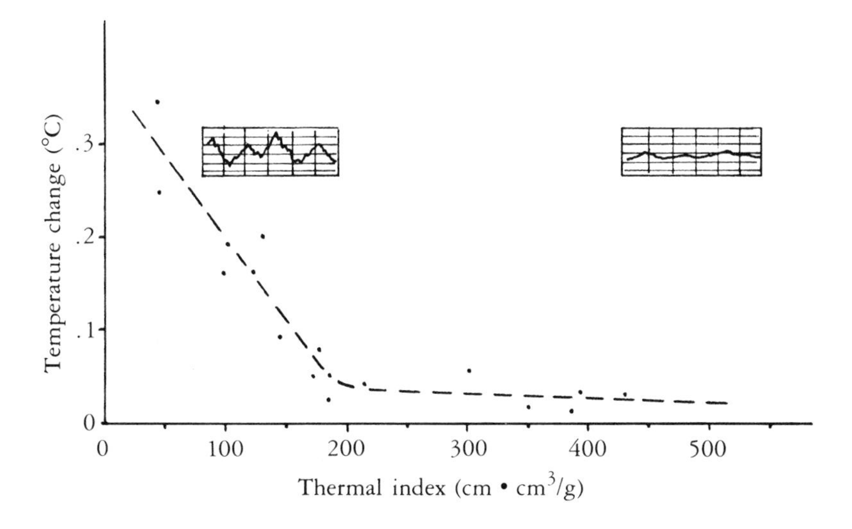

Fig. 5. Temperature fluctuations under snow in relation to snowpack thermal index values. The y-axis is the ratio of 24-hour subnivean temperature change to 24-hour air temperature change. Insets at the top depict graphically the daily temperature changes under snow of greatly different thermal indices.

body-air temperature differences by remaining under the snowpack, a strategy that will be discussed in some detail later. But the effect of snow on the energy relations of both plants and animals goes much further than its role as an insulative blanket, for snow greatly alters the partitioning of energy from sources external to the snowpack. To understand fully the influence of snow on winter-active organisms, one must consider its role in the disposition of incoming solar energy and outgoing terrestrial radiation.

SNOW AND RADIANT ENERGY

The maximum amount of radiant energy received at the earth's surface on a clear summer day, at mid latitudes, is on the order of 3000 joules/day. As the winter solstice approaches, maximum insolation decreases to less than 1700 joules/day. However, this seasonal re-

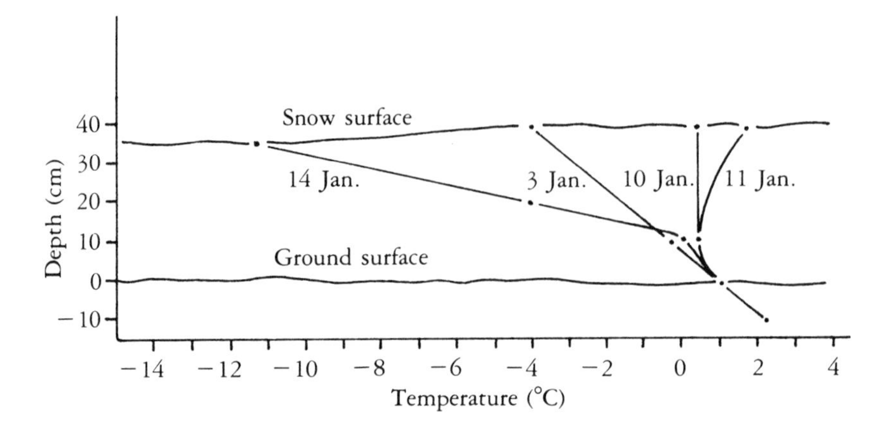

Fig. 6. Snowpack temperatures over a two-week period in mid winter. Air temperatures ranged from −35° C to just above freezing with a rain. To the small mammal active under the snow, however, ambient temperatures remained very stable and near 0° C—a situation typical for all but extreme northern latitudes, where subnivean temperatures may be lower.

duction in incoming solar energy tells only half the story. It is the fate of this radiation over snow-covered ground that is of greatest interest in our understanding of winter ecology.

As anyone who has spent a day on the snow in bright sunlight can attest, snow is highly reflective of incoming solar or short-wave radiation. Fresh snow is, in fact, nature's best reflector, turning back 75 to 95% of the sunlight striking its surface. This means that only a small portion of the total solar energy reaching the snow-covered ground is available to perform work in the snowpack (e.g. raise the temperature of the snow); the rest is reflected and lost back to space or absorbed by objects above the snowpack. As a snowpack ages, however, and the surface accumulates dust or bark and leaf litter, its reflectance may decrease to as low as 45%. The fate of the short-wave energy that does penetrate the snow surface will be discussed momentarily as it pertains to the light regime in the subnivean environment.

As good a reflector of short-wave radiation as snow is, it is nearly a perfect *absorber* of the long-wave or heat energy emitted by terrestrial objects. Physicists refer to an object that absorbs 100% of the energy incident upon it as a "black body," and nothing in nature so closely approximates a black body, with respect to energy in the longer wavelengths, as does a snowpack. Because every natural object emits some heat energy by virtue of its own molecular activity, the snowpack, then, is continually absorbing energy from its surroundings. The most visible evidence of this is the appearance, especially in the spring, of pronounced depressions in the snowpack around the trunks of trees, sometimes extending right to the ground surface. The tree absorbs incoming solar radiation, including some of that which the snow is reflecting; this absorption of energy increases molecular activity and raises the temperature of the absorbing surfaces; and the tree in turn emits more long-wave or heat energy. The snowpack absorbs this energy with great efficiency, which then increases the rate of sublimation around the tree trunk. The hollow is formed more by sublimation—the conversion of ice directly to vapor—than by actual melting of the snow (fig. 7).

Just as the snowpack continually absorbs heat energy, it also efficiently emits or radiates energy back to its surroundings. If our color vision extended beyond red light and into the longer wavelengths of heat energy that are invisible to us, we would see the snow surface emitting this "light" of long-wave energy. At night under a clear sky, when incoming radiation is greatly reduced, the snow surface in fact loses much more energy than it gains. Because of this energy loss the snow surface becomes colder, affecting in turn the air in contact with it, with the result that a cold layer of air often develops over the snowpack. This condition is referred to as a temperature inversion because it is a departure from the normal situation where the air close to the ground is warmer than the air aloft.

The practical point of this is that the lowest temperatures in any winter environment generally occur during the night right at the snow surface, provided there is no overhead obstruction to absorb

Fig. 7. The absorbing snowpack. Depressions in the snowpack around trees in this deciduous forest result from the absorption of long-wave or heat energy by the snowpack. The dark tree trunks absorb incoming solar radiation, the temperature of their absorbing surfaces is raised, and the trees then emit more energy at longer wavelengths. This energy is absorbed with near perfect efficiency by the snowpack, resulting in increased sublimation around the tree trunk. (Photo by Peter Marchand.)

outgoing radiation and reradiate it back to the snow. Temperatures then increase upward from the snow surface and downward into the snowpack. Figure 8 illustrates this for a northern forest clearing on a cloudless January night. It is not at all unusual to see such nighttime air temperature inversions develop over the snowpack and extend several meters above the snow if the air is calm (see box on "Ice Fog"). If you were looking for an overnight shelter under circumstances like this, you would be considerably more comfortable beneath the forest canopy where long-wave energy is being reradiated downward to the snow surface. For the same reason, snowmelt is often delayed the longest in small forest clearings where the surrounding trees intercept incoming solar radiation but allow the unimpeded loss of heat energy from the snow surface to space (fig. 9).

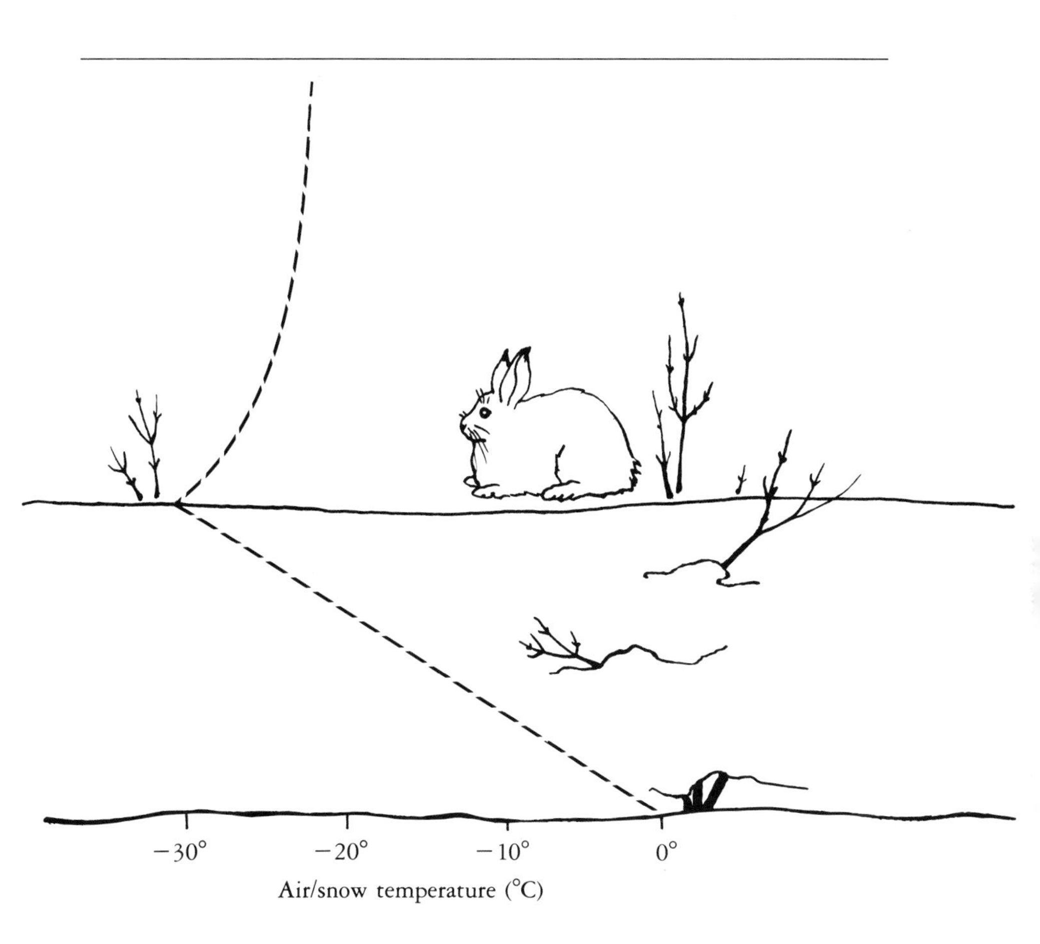

Fig. 8. Nighttime temperature profile over snow-covered ground. Air temperature is lowest on a clear night, right at the snow surface—in this case 7° colder than at a height of 2 m above the snow. The temperature increases rapidly beneath the snow surface.

ICE FOG

During the long nights in arctic regions, the snow-covered ground continually radiates heat to space, and air temperatures at the surface become increasingly colder. As long as the sky is clear and the air calm, temperature inversions continue to develop, with the height of the inversion sometimes extending several hundred meters above the surface.

a

b

Such strong temperature inversions have important implications with regard to air pollution in the North. While exhaust fumes from automobiles and emissions from smokestacks normally rise in the atmosphere because they are warmer and lighter than the air into which they are emitted, a strong temperature inversion with warmer air aloft prevents this from happening. Instead, the polluted air becomes trapped at ground level, resulting often in the formation of ice fog. At very low temperatures water vapor released from the combustion of fossil fuels freezes around other foreign matter emitted from the burned fuel to form a very dense fog consisting of minute ice particles suspended in the air. In the street scenes (*a* and *b*) photographed at midday in Barrow, Alaska, visibility in the ice fog is less than 30 m. (Photos by William Brower.)

This also explains why the coldest part of winter in the North does not coincide with the shortest days of the year—the time of lowest incoming solar radiation. It is because the snow-covered ground continues, beyond the winter solstice, to radiate more heat than it gains due to the high short-wave reflectance and efficient long-wave emittance of snow. The annual distribution of available solar energy at polar latitudes is actually asymmetrical with respect to total incoming radiation because of the presence of a snowcover that lasts well into spring. If we measure the individual components of energy exchange and subtract all the outgoing terrestrial radiation from the absorbed incoming radiation (total incoming minus reflected), we find for latitudes north of the Arctic Circle and south of the Antarctic Circle an energy deficit that persists well into the spring (fig. 10). This energy deficit is alleviated only by the occasional influx of warm air masses into these regions from temperate or tropical latitudes.

Returning to the disposition of incoming solar radiation, it is of

Fig. 9. Delayed snowmelt in forest clearings. In small forest openings, where the height of the surrounding trees is greater than the diameter of the clearing, incoming solar radiation is intercepted by the trees and very little is received directly at the snow surface. At the same time, loss of outgoing long-wave or heat energy from the snowpack is unimpeded due to the lack of overhead obstruction. The result is a negative energy balance (the snow surface loses more energy than it gains), which delays snowmelt. Increased deposition of snow in the forest opening from the crowns of nearby trees also contributes to later lying snowpacks in these situations.

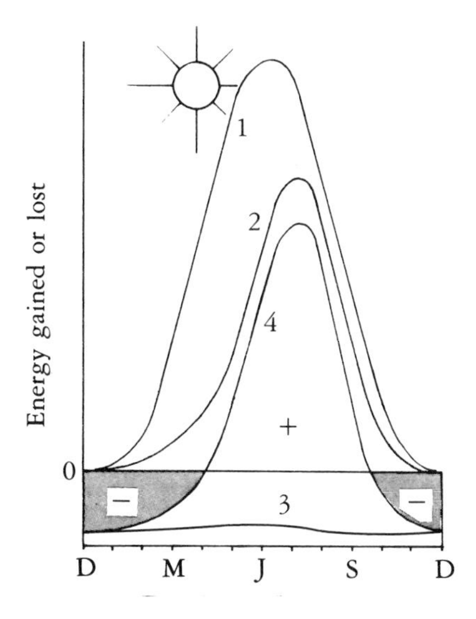

Fig. 10. Annual radiation balance at high latitudes. Curve 1 *indicates the total amount of solar radiation received per month at the earth's surface, at a latitude of 70°N. Because of the persistence of a highly reflective snowcover, only a relatively small proportion of this is absorbed (*Curve 2*), until June when the ground is finally snow-free. For most of the winter, radiant energy emitted from the snow surface (*curve 3*) is greater than the total absorbed, giving a negative energy balance (*curve 4*), as indicated by the shaded areas.*

interest now to ask what happens to the light that does penetrate the snow surface and how might its transmission through the snowpack be affected by seasonal changes in snow depth and density. Though we may be dealing with only a very small amount of radiant energy under the snowpack, its biological significance may be disproportionately high. Even very low light levels under the snow may influence a number of life processes throughout the winter. We will be considering evidence later, for example, that plants may be able to utilize light energy penetrating deep snowcover for chlorophyll production and photosynthesis. The light regime under snow may also affect small mammal population dynamics by directly influencing sexual maturation and reproductive behavior of small mammals in the subnivean environment. It is important, therefore, to have some idea of how light behaves as it passes through snow.

If snow at .2 g/cm^3 is thought of as representing one end of a density gradient, with glacier ice (.7 g/cm^3) somewhere in the middle and water (1 g/cm^3) at the other end, then one might conclude from experience that the transparency of snow increases as its density increases to that approaching ice. This has been the traditional view and there is much evidence in the literature supporting it, particularly from researchers interested in how light affects the growth of snow algae on alpine glaciers. However, if we consider our experience with snow more closely, a problem arises with this generalization. A snowball in the hand or a slab of wind-packed snow certainly appears to transmit less light (is more opaque) than a handful of loose snow held to the light, even though the density of the snowball or wind slab may be two to four times greater than the loose snow. But if our eyes are not deceiving us, how can we reconcile this observation with apparent discrepancies in the literature? The answer lies in the fact that light penetrating the snowpack behaves very differently under opposite extremes of snow density. Whether light transmission increases or decreases as the snowpack becomes more dense actually depends on whether we are dealing with a seasonal snowpack of relatively low density, typical of northern forest regions, or the highly metamorphosed permanent snowfields of arctic and alpine regions. More specifically, the behavior of light in snow depends on the degree to which the snowpack has been altered by the metamorphic processes described earlier.

The accumulation and settling of fresh snow is accompanied by a marked decrease in light transmission. This is illustrated in figure 11 for snow of two different densities. At an average density of .21 g/cm^3, just 3% of the light that passes through the surface penetrates a depth of 20 cm. Less than two-tenths of a pecent is transmitted to 40 cm. A slight increase in density to .26 g/cm^3 reduces the amount of light passing through the snow by 50% or more at any given depth.[1] This means that an increase in density of only .05

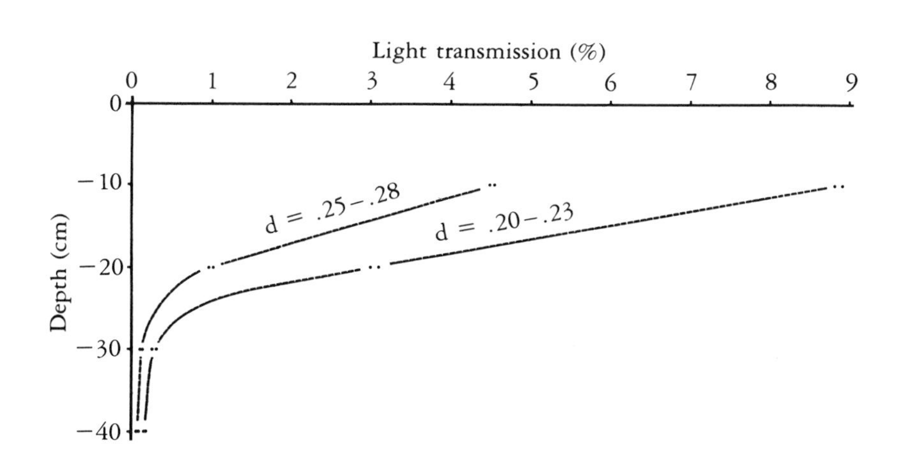

Fig. 11. Absorption of light in the snowpack. Light transmission (percent of visible light penetrating the surface) is shown here as a function of depth in snow of different densities. (From P. J. Marchand, "Light Extinction under a Changing Snowcover," in Winter Ecology of Small Mammals, *ed. J. F. Merritt, Carnegie Museum of Natural History Spec. Publ. 10 (Pittsburgh, 1984), 33–37.)*

g/cm^3 in the surface layer would have approximately the same effect as the addition of 10 cm of new snow.

The sharp reduction in light transmission after a snowfall is most pronounced at low densities. Beyond .3 g/cm^3 the effect of increasing density diminishes rapidly, and at .5 g/cm^3, maximum light absorption by the snowpack is reached. At this point the light transmission curve bends upward again, with further increases in density then resulting in an increase in light transmission as had been previously reported.[2] This turnaround in light transmission is illustrated in figure 12 for a range of snow densities from .1 to .7 g/cm^3.

It is of interest to note that the density of maximum light absorption (.5 g/cm^3) is sometimes referred to as the "critical density" of snow. This is considered to be the maximum snow density attainable by compaction alone. In order to attain higher densities, a sintering or coalescence of ice grains must occur—an effect that may

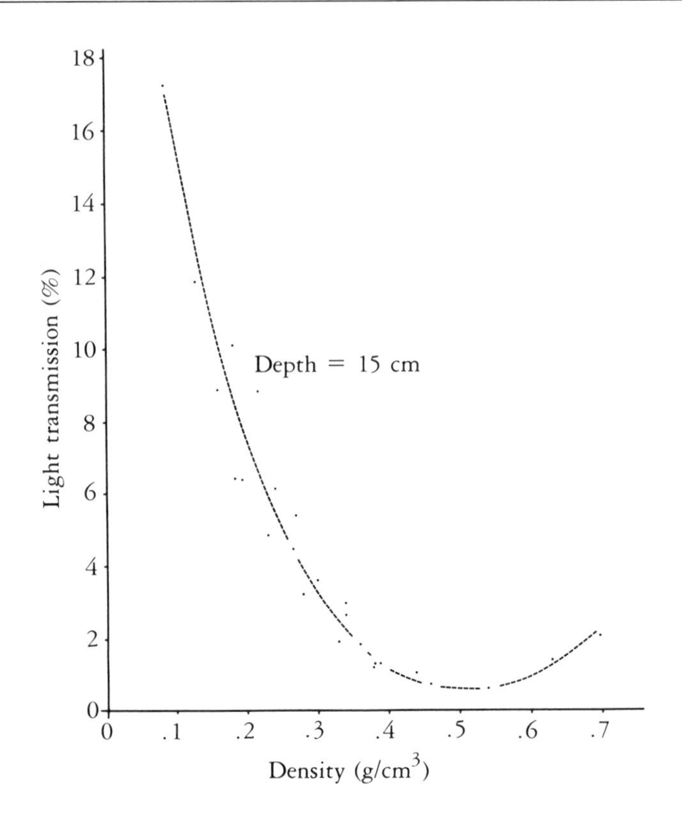

Fig. 12. Effect of snowpack metamorphism on light penetration. Light transmission (percent of visible light penetrating the surface) is shown here in relation to snow density. (From Marchand, "Light Extinction," 35–37.)

be achieved experimentally by warming the snow under pressure or adding water in a way similar to the processes occurring naturally over time in permanent snowfields or glaciers.

The changes in snowpack structure that take place above the critical density of .5 g/cm^3 may well explain why light transmission begins to increase at this point. Every time a light beam passes through and out of an ice grain, its direction is changed by refraction. The bending of the light beam actually occurs right at the ice-air interface. Therefore, the greater the total surface area of ice grains in a given volume of snow, the greater the amount of internal light scattering caused by refraction, and the greater the opportunity for absorption of light by the ice molecules. In the process of destructive metamorphism, increases in density occur primarily as a result of a closer packing of small grains—and more grains per unit volume means more total surface area. This increases refraction and scattering, and subsequent absorption then decreases the amount of light transmitted through the snowpack. In contrast, densities above .5 g/cm^3 are attained primarily through an increase in overall grain size due to the growth of bonds between individual ice particles. This leads to a reduction in total surface area which, in turn, reduces internal refraction and increases light transmission to greater depths (fig. 13).

How, then, is the light regime of a subnivean environment altered as the snowpack undergoes seasonal changes in depth and density? As already discussed, new fallen snow undergoes a very rapid change (destructive or equi-temperature metamorphism) as thermodynamically unstable crystals are transformed into aggregates of rounded ice grains. Within a few hours snow density may increase from less than .1 g/cm^3 to nearly .2 g/cm^3. This is followed by settling, often aided by wind, which then reduces the size of snowpack pore spaces and further increases snow density. Light transmission, therefore, will decrease sharply with time after a snowfall.

As soon as a temperature gradient is established in the snowpack, upward-moving water vapor causes a dissipation of ice crystals in the bottom layers and a growth of grains in the upper, colder layers of the snowpack (constructive metamorphism). This alone would tend to offset the decrease in light transmission, but seasonal snow accumulation and packing result in a continued increase in density through destructive metamorphism to values approaching .4 g/cm^3. A plot of light penetration with time would, thus, be asymmetrical with respect to seasonal changes in both snow-cover and incoming solar radiation (fig. 14) because the increase in incident radiation through the winter is offset by steadily increasing snow density. Less than .1% of incident light may reach the ground surface from mid January to early April in areas where snow depth is

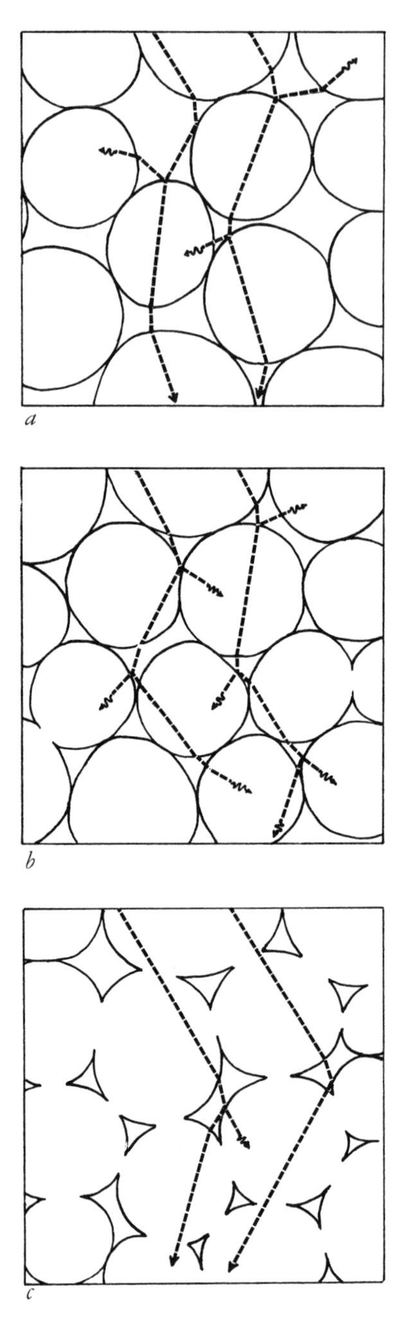

Fig. 13. Light scattering in the snowpack. A light beam passing through a snow crystal refracts or bends at the ice-air interface, resulting in a scattering of light within the snowpack (a). *An increase in snowpack density caused by closer packing of ice grains creates more surfaces to scatter light in a given volume of space and, therefore, increases opportunity for absorption of light* (b). *At still higher densities, however, ice grains coalesce, reducing the amount of surface area per unit volume, and light passes through with less scattering and absorption* (c). *This explains the turnaround in light transmission as the snowpack undergoes metamorphism.*

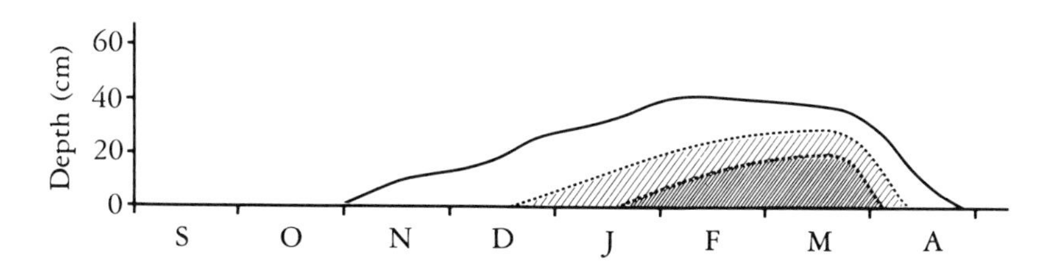

Fig. 14. Changing light penetration with seasonal changes in snow depth, density, and solar radiation (44°N latitude) for a hypothetical snowpack. Light shading represents < 1.0% of incident radiation and dark shading < 0.1%. Average snowpack densities of .21, .26, and .3 g/cm³ for January, February, and March, respectively, were used for calculations, and a constant snowpack reflectance of 75% was assumed. (From Marchand, "Light Extinction," 33–37.)

40 cm or more and average density exceeds .25 g/cm³.

Before leaving this discussion on the nature of light under snow, one other aspect of its behavior should be dealt with, and that is its quality. So far we have been dealing in a general way with "white light" or visible radiation from the sun without considering what colors or wavelengths of visible light penetrate snow most effectively. Among the best studies to date on the quality of light penetrating snow are those of Curl et al.[3] and Richardson and Salisbury.[4] Curl and his colleagues were interested in light energy as it affected the growth of algae in alpine snowfields. Using a spectroradiometer to measure light penetration of specific wavelengths under various snowpack conditions, they found that light absorption by snow was most efficient in the red region of the spectrum, with blue light reaching greater depths. A generalized graph of light penetration by wavelength from their data is given in figure 15. Richardson and Salisbury, who were also interested in plant response to light, constructed a small laboratory beneath the snowpack and used a photometer equipped with a photomultiplier tube and interference filter (which enabled selection of specific wavelengths) to measure extremely low levels of light under almost 2 m of snow. Their studies indicate that the wavelength of maximum penetration is centered

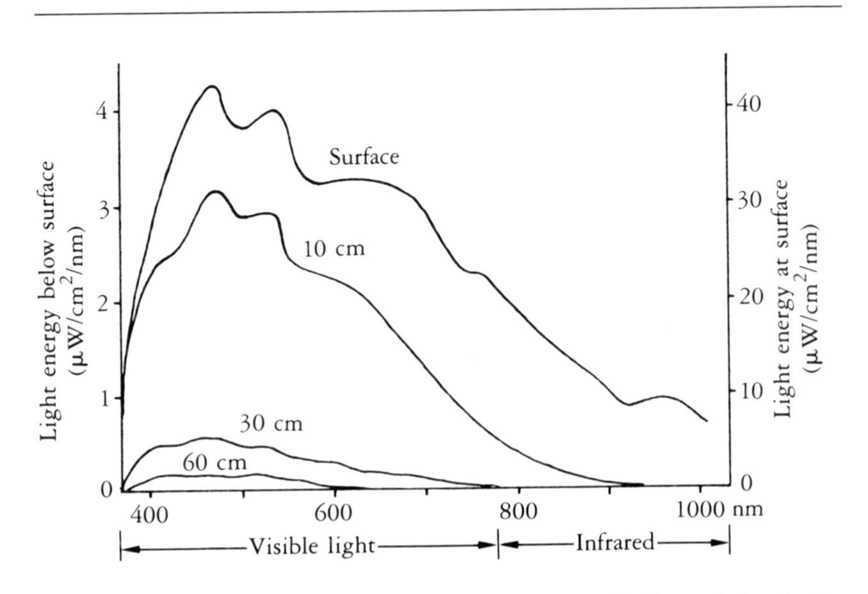

Fig. 15. Light penetration by wavelength under snow of different depths. During mid winter, snow absorbs red light most effectively, with blue and blue-green light reaching the greatest depths. (Redrawn from H. Curl, J. T. Hardy, and R. Ellermeier, "Spectral Absorption of Solar Radiation in Alpine Snowfields," Ecology *53 (1972), 1189–94.)*

around that of blue-green light (about 500 μm). However, as the snowpack ages, its reflectance decreases, and as solar angles become higher, the proportion of red light penetrating the snowpack, relative to blue light, increases.[5]

Though we now know that blue or blue-green light penetrates to considerable depths in snow, we have also seen that absolute light levels under 40 or 50 cm of snow are very low at typical midwinter snow densities. Any way we measure it, it is dark down there. But that is only *our* perception. What of the organisms surviving winter under the snow? The next chapter considers whether there may, in fact, be sufficient light energy available to support some plant activity in the subnivean environment. The response of animals to such low light levels, however, is unknown and raises several interesting questions: To what extent might social interactions of subnivean mammals be influenced by changing light levels under the snow? Does light quality play a role in sexual maturation and reproductive activity in the subnivean environment, or do mammals perceive only total darkness under 40 or 50 cm of snow? Do daily (circadian) rhythms drift out of phase with "external" time-setters throughout the winter, or are the low light levels under snow sufficient to regulate daily activity patterns in subnivean mammals? These questions remain a challenge to animal ecologists. Clearly, the response of small mammals to the light conditions under snow is an area in need of much more study before any generalizations may be drawn.

3 PLANTS AND THE WINTER ENVIRONMENT

In the introductory section of this book, we considered a number of alternatives for overwintering success in the North. Our attention then was focused mainly on animals, perhaps leaving the impression that there were few evolutionary options for plants. This is not entirely the case. Some plant species have evolved mechanisms for avoiding the rigors of the winter season—for example, the overwintering seed in the case of annuals, or the below-ground corm or regenerative root stock in the case of many herbaceous perennials. Others have evolved morphological and physiological adaptations for resisting the stresses of winter. Certainly the deciduous habit of northern and midlatitude hardwoods can be considered an adaptive advantage in reducing water loss during winter when water supply is greatly limited. In the case of aspen in northern forest regions, we might even speculate that its high chlorophyll concentration in bark tissues is of adaptive value, compensating in part for the long leafless season—a possibility that will be discussed in more detail later. And the more pronounced spire form of conifers in the Far North, a result of hormonal suppression of lateral branch growth, is viewed by some as being of adaptive advantage in heavy snow country.

As was the case with animals, there are no "best answers" in an evolutionary sense for overwintering success in plants. The coniferous growth form seems overwhelmingly favored throughout the vast boreal forests of the northern hemisphere, yet there are a number of deciduous broadleaf tree species that grow right to the arctic timberline. Larch (*Larix* spp.), the deciduous conifer that combines the best of both growth forms, is capable of surviving the coldest winters on earth outside of Antarctica. Dahurian larch (*Larix dahurica*) forms the northernmost forest stands in the world, surviving in northeastern Siberia where January temperatures may go as low as −65° C. In the end it is the physiology of these plants, rather than their morphological differences, that gets them through the winter. Though the snowpack offers some advantage to plants that remain under cover, protecting them from many winter stresses (and subjecting them to one or two others, as we will see later), the survival of those plants perennially exposed during winter is largely independent of the presence of snow. For these plants the major problems of winter are related primarily to low temperature stress and desiccation (the two are not entirely unrelated); the one common denominator among all trees and shrubs of northern areas is some ability to withstand these two stresses.

ACCLIMATING TO THE COLD

Acclimation is the process by which plants each year become tolerant to subfreezing temperatures without sustaining injury. It is not an avoidance mechanism whereby the plant somehow manages to escape freezing (the plant produces no "antifreeze" as such). Rather, it is a way in which the plant becomes increasingly hardy or resistant to the potentially damaging effects of tissue freezing and ice formation. In order to understand fully the process of acclimation, it is helpful first to have some understanding of how freezing occurs in the plant and what happens at the lower threshold of cold tolerance that causes permanent injury. In the following paragraphs we will consider the sequence of events that occur at the cellular level when a woody plant tissue undergoes slow cooling across the freezing point of cell sap.

Imagine an experimental setup which closely monitors the freezing of a section of woody stem, a maple twig for example, as it is cooled. In this experiment the stem section has two fine wire thermocouples attached to it, one embedded under the bark to measure tissue temperature and the other mounted about a halfcentimeter away from the bark to monitor simultaneously air temperature. With these thermocouples we can plot cooling curves for both air and stem as we subject our sample to steadily decreasing temperatures. On the basis of numerous studies of freezing in woody plant tissues, this is what we would expect to see:

At first the decrease in stem temperature closely parallels air temperature. Both descend across the freezing point of water with no divergence of the cooling curves. At a temperature of perhaps $-5°$

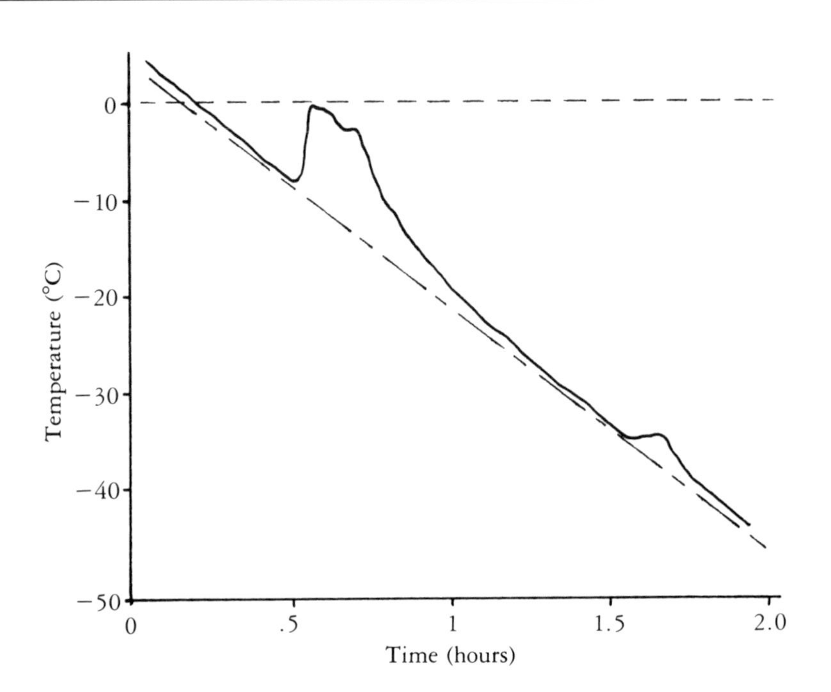

*Fig. 16. Typical freezing curve for a woody stem, in this case beech (*Fagus grandifolia*), undergoing slow cooling. The first exotherm, or rise in stem temperature, results from the release of latent heat as "flash freezing" occurs throughout the stem. The double peak indicates ice formation first in the water conducting xylem cells and large intercellular spaces and then within the cell walls and micropores of the stem. The second exotherm, occurring at a much lower temperature, coincides closely with the point of cell death and may represent intracellular freezing.*

or −8° C, however, we see a dramatic departure of the two curves (fig. 16). Air temperature continues to decrease at our prescribed rate, but now the stem shows an abrupt increase to just under 0° C. Because of the various attractive forces with which water is held in plant tissues, the tissue is able to "supercool"—that is, cool below its actual freezing point without forming ice. As we continued to lower the temperature, though, we eventually promote ice crystallization. In this supercooled state, the initial formation of ice starts a chain reaction of "flash freezing" throughout the tissue, resulting in the release of a measurable quantity of latent heat (an "exothermic" reaction). The accompanying rise on our cooling curve is referred to as the first freezing exotherm.

All the evidence to date indicates that this first exotherm represents ice formation in pore spaces outside the living cell, a reasonable conclusion since the concentration of dissolved substances (which lowers the freezing point of water) in the intercellular spaces of the plant is lower than in the cytoplasm of the cell. The actual freezing point of this intercellular water is generally no more than one- or two-tenths of a degree below that of pure water. If our experiment is done carefully enough, two exotherms associated with this first freezing event may appear. These are separated very closely in both time and temperature, with the first representing the freezing of water in the nonliving xylem cells (the water conducting cells) and large intercellular spaces of the stem, and the second indicating freezing in micropores and cell walls.[1] The initial formation of ice outside the living plant cell is vitally important, for the process that follows this forestalls intracellular ice formation, an event that is almost invariably fatal to the cell.

Because the vibrational energy of a water molecule on the surface of an ice crystal is much reduced, an energy gradient exists between unfrozen water in the cell cytoplasm and frozen water outside the cell. As a result of this energy difference, the more active liquid molecules tend to migrate toward the ice surface. This results in an orderly movement of cytoplasmic water through the plasma membrane and out of the cell, adding its mass to the growing intercellular ice crystal. This loss of water from the cell has the effect of increasing the solute concentration of the cytoplasm and thereby progressively lowering its freezing point. As we shall see later, an important aspect of the acclimation process is an increase in cell membrane permeability so that this outward migration of water is little impeded.

As we continue our cooling experiment, the progressively smaller amounts of water freezing onto the extracellular ice crystal contribute less and less latent heat and gradually our stem sample comes close to equilibrium with air temperature again. If we push our tissue sample to the limit of its tolerance, however, we may eventually (though not always) see a second, smaller exotherm that corresponds closely with the timing of cell death.

Just what happens at this second exotherm and point of cell death is not absolutely clear, but there are at least two hypotheses that deserve consideration here. One idea put forth more than a decade ago by Soviet scientists bears merit, though it remains unconfirmed by experimental data. This hypothesis holds that at some critical low temperature an abrupt decrease in plasma membrane permeability occurs that prohibits any additional water from migrating out of the cell. The entrapped water then supercools, followed by intracellular ice nucleation and the rapid propagation of ice throughout the cell, causing the exotherm. The implication here is that the lethal damage is of a mechanical nature, with cell membranes being the primary site of injury.

Whereas there has been no confirmation yet of a sudden change in plasma membrane permeability at some low temperature, there is accumulating evidence that freezing injury to living cells is in some way associated with an alteration in resilience of the plasma membrane that may occur during extreme contraction of the membrane.[2] It is believed that the intrinsic proteins bridging the two sides of the membrane (fig. 17) may be the actual sites of damage and that once these break down they can serve as channels for the leakage of potassium ions and sugars that usually accompanies freezing injury.[3] The rearrangement of some membrane lipids into nonlayered structures upon freezing and thawing may also alter permeability of the membrane.[4]

An alternative explanation for cell death at the second freezing exotherm is the "vital water" hypothesis of Weiser.[5] This idea holds that, during freezing, a point is reached where all readily available water has moved out of the cell and is frozen intercellularly, and only water bound to structural macromolecules remains in the protoplasm. With a continued decrease in temperature, removal of

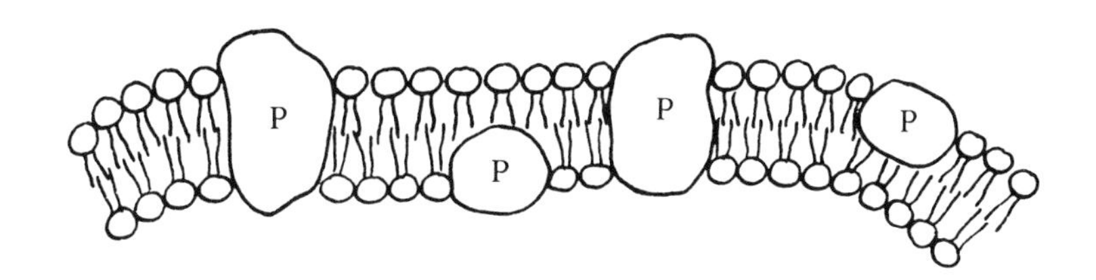

Fig. 17. Membrane failure with cell freezing. Current models show membranes as a double layer of lipids interspersed with protein macromolecules. Some investigators believe that the intrinsic proteins bridging the two sides of the membrane are the sites of freezing injury, resulting in the leakage of cell electrolytes. Another line of thought suggests that the rearrangement of some membrane lipids into nonlayered structures upon freezing and thawing may lead to membrane failure.

bound water initiates a chain reaction of denaturation (probably of membrane proteins among other things), additional water loss, and death by irreversible dehydration or the accumulation of toxic concentrations of solutes—the latter also capable of altering membrane properties.[6]

Whichever hypothesis proves correct, there is ample evidence that it is ice formation and not low temperature per se that directly or indirectly causes freezing injury. Even nonhardy plant tissues can survive the $-196°$ C temperature of liquid nitrogen if they are cooled slowly before immersion. This is because immersion in liquid nitrogen solidifies water without molecularly reorienting the water into an ice crystal, a process called vitrification. Just recently it has been reported that such amorphous solidification of cell contents (termed "glass formation") can occur in nature at surprisingly high subfreezing temperatures and without release of heat in a detectable exotherm. In balsam poplar (*Populus balsamifera*), intracellular fluid was observed to solidify, with slow cooling, at temperatures below $-28°$ C, effectively preventing cell injury by either intracellular ice formation or dehydration.[7] This is the first time that high temperature glass formation in a plant has ever been reported. Its possible occurrence in other species may explain why a second exotherm has

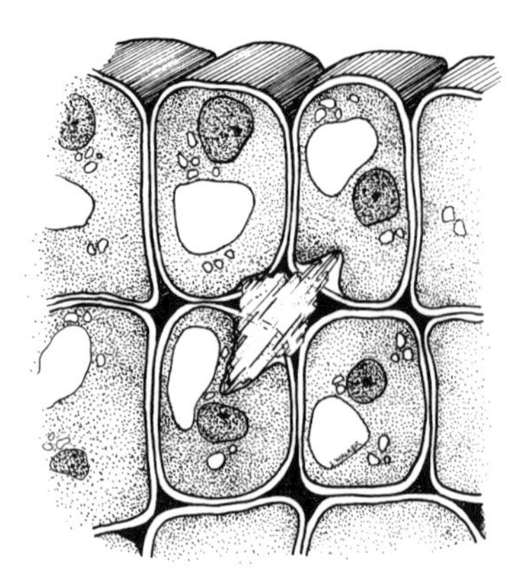

Fig. 18. Ice formation in living plant tissues. Ice crystals growing in intercellular spaces as water moves out of surrounding cells easily penetrate the fabric of the cell wall. The plasma membrane, though, remains resistant to penetration and merely deflects inward, thus preserving the integrity of the living cell.

never been seen in some winter acclimated trees and why their resistance is sometimes so extraordinary. In any event, ice outside the cell is detrimental only in the extreme case hypothesized by Weiser, where it may draw so much water out of the cell that irreversible damage occurs. Otherwise, a growing extracellular ice crystal can penetrate the fabric of the cell wall without harm, merely deflecting the plasma membrane as the membrane itself contracts with diminishing protoplasmic volume (fig. 18). This often observed phenomenon is referred to as "frost plasmolysis" and is harmless to the tissue as long as intracellular freezing does not occur.

It is important here to distinguish between slow cooling, at rates on the order of 10° C per hour that frequently occur in nature, and rapid freezing that can be accomplished in the laboratory and that occasionally occurs in nature. The difference has much to do with the survival of even the hardiest of plant species. If cooling is rapid enough that withdrawal of water from the cell cannot keep pace with the rate of freezing, then ice formation may occur within the cell and this is invariably fatal. In a study of winter damage to northern white-cedar (*Thuja occidentalis*) in Minnesota, White and Weiser found that at sunset, where the horizon was abrupt, leaf temperatures on the sunlit southwest side of the tree dropped 9.5° per minute across the freezing point of cell water.[8] This rapid freezing apparently trapped water inside the cell, where it froze and resulted in cell death. Under slower freezing rates this species was capable of withstanding temperatures of −87° C without damage. It may be that much winter injury to evergreens that has been attributed to desiccation is the result of rapid freezing instead. Since freezing damage so closely resembles desiccation damage in mechanism, it is often impossible to tell the two apart by appearance.[9]

Having some understanding now of the freezing process in woody plants, it is appropriate here to ask how plants acclimate or acquire freezing resistence. This is by no means a simple question. We know from experience that the effect of low temperature varies considerably among plants; even different populations of the same species may vary in their ability to withstand freezing. We know that a light frost during the growing season may have a far more damaging effect on exposed tissue than the coldest temperatures of winter. And we know that individual plants can sometimes, by proper conditioning, be made more or less resistant to freezing damage over time. In addition to these variables we find that different tissues of the same plant may have very different thresholds of injury, and that these tolerances can change, depending on the exposure of the tissue. It is indeed a very complex situation and many questions remain unanswered today.

In spite of what remains to be learned, there appear to be some constants in the acclimation process. Freezing acclimation, the acquiring of some ability to withstand subfreezing temperatures without cell injury, begins in late summer and is ordinarily a two-step process. If you collect twig samples of a given tree species on a weekly basis, beginning in the latter part of the growing season, and subject them to a range of low temperatures in order to determine their threshold for survival, you will likely find at first that freezing injury occurs at temperatures very close to 0° C. Then, as the growing season comes to an end, you will begin to see an increase in their survival of subfreezing temperatures. The temperature at which freezing injury occurs will slowly drop and it will appear as if the tree will soon be able to withstand the temperatures

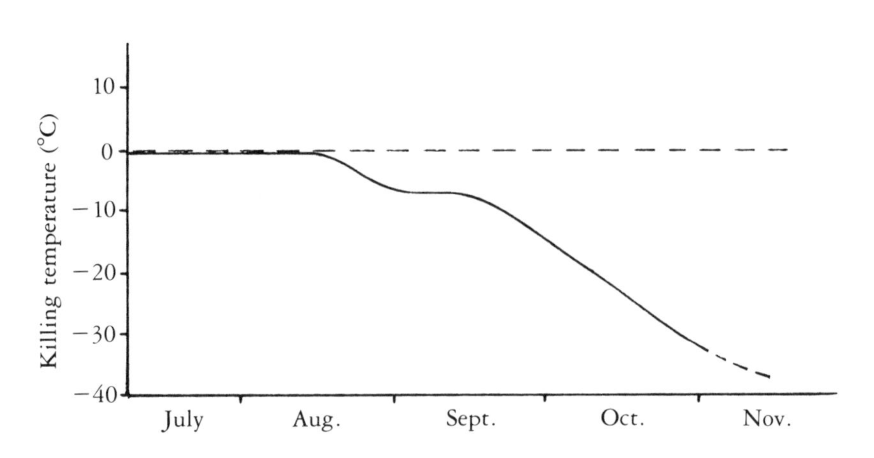

Fig. 19. Woody plant acclimation to the cold. Freezing acclimation in woody plants normally occurs as a two-phase process. The first phase is induced by decreasing day length and results in sufficient resistance to withstand the first light frosts of autumn. Exposure to freezing temperatures then induces a second phase of acclimation that leads to deep-winter resistance typical of the species. (From C. J. Weiser, "Cold Resistance and Injury in Woody Plants," Science *169 (1970): 1269–78.)*

of midwinter. However, as you continue to monitor the acclimation process, you will discover, after only two or three weeks perhaps, that the resistance of the plant reaches a plateau (fig. 19). For a while now the minimum survival temperature remains steady at around −5° to −10° C and it looks as if that is all the freezing resistance the tree will acquire. With time, however, the minimum survival temperature begins to decrease again as the acclimation process enters its second phase. This phase seems to be induced by the first hard frosts of the fall season, and this time the plant will reach its full midwinter freezing resistance.

The first stage of acclimation is apparently linked to growth cessation, influenced primarily by decreasing day length, but induced also by water stress. For example, the bristlecone pine attains a greater degree of freezing resistance early in the fall if active growth has been prohibited by drought during the latter part of the summer.[10] Wheat seedlings subjected to water stress for two weeks, which limits growth, and then placed in darkness have been found to develop some freezing tolerance.[11] Low but positive temperatures (e.g., 2° to 5° C) may also induce this stage of acclimation in herbaceous plants, though again the response may be related to growth cessation rather than to temperature itself.

Regardless of whether the induction factor is related to decreasing daylight, low temperature, or some other seasonal change, it is clear that the timing of phase I acclimation is subject to some preconditioning, though not all populations of a given species respond in a similar manner to the same cues. This was shown in a number of experiments by Weiser and his students using regional populations of red-osier dogwood (*Cornus stolonifera*) from several different climatic zones, which were transplanted and grown in St. Paul, Minnesota.[12] Populations from northern interior regions that normally experience very cold winters entered the first stage of acclimation as much as a month earlier than populations from milder coastal areas, though all populations eventually attained the same degree of freezing resistance.

The onset of freezing acclimation seems to involve the translocation of some biochemical compound, perhaps a simple sugar or specific hormone, which comes from the leaves, the site of perception of short days. There are a number of pieces of evidence that support this translocation hypothesis, among which are the observations that in herbaceous plants, this stage of hardiness can occur in darkness only if the plants are cold treated and supplied with sucrose; and in woody plants the leaves of hardy genetic strains can enhance the acclimation of nonhardy strains when the two are grafted together.[13] In addition, plants that are severely depleted in photosynthetic reserves do not acclimate.

The exact identity of this translocated hardiness-promoting factor remains elusive, perhaps because there may be more than one mechanism involved in the induction of freezing resistance. The plant hormone abcisic acid (ABA) almost surely plays a role in the acclimation process, at least indirectly. ABA is a photoactive hormone that is synthesized primarily in leaves and increases in concentration as day length decreases. ABA concentration also increases dramatically with even mild water stress. ABA is a nearly universal growth inhibitor, a prerequisite to acclimation, and is known to increase the permeability of membranes to water, an all-important factor in freezing resistance.[14]

In addition to the increasing ABA concentrations at the end of the growing season, a great number of other biochemical changes are associated with the acclimation process—enough, as others have pointed out, to support almost any hypothesis for freezing resistance ever proposed. Among them, changes in specific sugars and proteins, along with increasing lipid unsaturation, appear to be linked with low but positive temperature. Associated with the second stage of acclimation, induced by subfreezing temperatures, is a rapid alteration of membranes that may also involve a reorientation of macromolecules into more stable forms that can resist dehydration. For example, proteins may be converted from a polymerized configuration (many units linked together, having a high combined molecular weight) to a depolymerized state in which water is bound to each protein unit.[15] Alternatively, sugars may replace water, forming protective shells around sensitive proteins.[16]

Seasonal alteration of cellular membranes during acclimation and deacclimation is an important process both with respect to changing membrane permeability and structural integrity. In addition to the possibility that membrane proteins undergo some changes, there is ample evidence now that the lipids of at least thylakoid (chloroplast) membranes undergo pronounced seasonal alteration, too. A substantial increase in total lipid content in different age classes of Norway spruce needles closely parallels seasonal changes in ambient temperature and plant freezing resistance. Similar correlations of lipid content and freezing resistance have been reported for other herbaceous and woody species.[17] Of equal importance is the marked decrease in saturation of membrane lipids, which apparently shifts the crystallization point of the lipids to a lower temperature.[18] This has been observed in other species and may be important in maintaining higher membrane flexibility during cold periods, thus minimizing the possibility of structural damage at low temperature.[19]

It is clear that acclimation is an active process, not simply a consequence of growth cessation, and that among the many metabolic changes taking place at this time, the change in membrane permeability is of paramount importance in allowing water withdrawal and preventing intracellular ice formation. Scientists believe, in fact, that in the winter-hardy plant, the plasma membrane offers virtually no resistance to the movement of water. If this is the case, then the osmotically active substances such as sugars, organic acids, and water-soluble proteins, all of which increase in concentration during acclimation, may help to counteract the secondary effects of increased membrane permeability and protect the cell against lethal desiccation. However, the mechanism by which some species, or even different tissues of the same plant, become much more resistant than others is not known.

Even in its final state of freezing resistance, the plant may remain sensitive to fluctuations in ambient temperature. A midwinter cold wave, for example, may stimulate an increase in freezing tolerance by several degrees in as short a time as five to ten days.[20] Some researchers even recognize a third phase of acclimation that may be induced by prolonged exposure to temperatures on the order of −30° to −50° C and may result in a level of freezing tolerance not ordinarily attained in many species.[21] Dehardening in response to high temperatures (e.g., 15° C) can occur even more quickly, though rarely are midwinter thaws warm enough or prolonged enough to cause serious loss of resistance in hardy species. By the few accounts in the literature, this responsive adjustment to normal temperature fluctuations in winter may cause a shift in minimum freezing tolerance on the order of 15° to 20°.[22]

Freezing injury no doubt has been a strong selective pressure in the evolution of many plant populations. Some tree species exhibit a low temperature tolerance so close to the average minimum temperature at their northern range limit (table 2) that it is safe to suggest freezing injury as the limiting factor in their distribution.

Table 2. Tree species whose freezing tolerance closely matches the minimum temperature at their northern range limit

Species	*Killing temperature (°C)*
Live oak, *Quercus virginiana*[a]	−8
Pacific bayberry, *Myrica Californica*	−10
Redwood, *Sequoia sempervirens*	−15
Oregon white oak, *Quercus garryana*	−15 to −20
Southern magnolia, *Magnolia grandiflora*	−15 to −20
Slash pine, *Pinus elliottii*	−10 to −20
Swamp chestnut oak, *Quercus michauxi*	−20
Sweetgum, *Liquidambar styraciflua*	−25 to −30
Eastern redbud, *Cercis canadensis*[b]	−35 to −40
Northern red oak, *Quercus rubra*	−40 to −41
American beech, *Fagus grandifolia*	−41
Black cherry, *Prunus serotina*	−42 to −43
Sugar maple, *Acer saccharum*	−42 to −43
American hornbeam, *Carpinus caroliniana*	−42
White ash, *Fraxinus americana*	−42 to −46
Shagbark hickory, *Carya ovata*	−43 to −46
Eastern hophornbeam, *Ostrya virginiana*	−41 to −48
Yellow birch, *Betula alleghaniensis*	−44 to −45

[a]A. Sakai and C. J. Weiser, "Freezing Resistance in Trees in North America with Reference to Tree Regions," *Ecology* 54 (1973): 118–26.
[b]Freezing injury in stem sections reported in M. F. George, M. J. Burke, H. M. Pellet, and A. G. Johnson, "Low Temperature Exotherms and Woody Plant Distribution," *HortScience* 9 (1974): 519–22.

Yet other species exhibit freezing tolerance way beyond any low temperature that they have or will likely ever experience (table 3), and one wonders by what mechanism they have evolved such extreme tolerance. Between these two groups there are others that show re-

Table 3. Tree species whose freezing tolerance greatly exceeds the lowest temperature within its range (or species that show little variation in tolerance over a wide geographical area)

Species	*Killing temperature (°C)*
Baldcypress, *Taxodium distichum*[a]	−30
Black cottonwood, *Populus trichocarpa*	−60
Eastern cottonwood, *Populus deltoides*	−50 to −80
Eastern hemlock, *Tsuga canadensis*	−60
Black willow, *Salix nigra*	−60 to −80
Basswood, *Tilia americana*	−80[b]
Northern white cedar, *Thuja occidentalis*	−80 ↓
Red pine, *Pinus resinosa*	−80 ↓
Jack pine, *Pinus banksiana*	−80 ↓
White spruce, *Picea glauca*	−80 ↓
Black spruce, *Picea mariana*	−80 ↓
Larch, *Larix laricina*	−80 ↓
Balsam fir, *Abies balsamea*	−80 ↓
Balsam poplar, *Populus balsamifera*	−80 ↓
Quaking aspen, *Populus tremuloides*	−80 ↓
Paper birch, *Betula papyrifera*	−80 ↓

[a]Sakai and Weiser, "Freezing Resistance of Trees," 118–26.
[b]George et al., "Low Temperature Exotherms," 519–22.

markable genetic flexibility in adjusting freezing tolerance to the minimum temperatures typical of the region in which they grow (table 4). Douglas fir growing on the coast of Oregon, where winters are relatively mild, attain a freezing tolerance to only −20° C, but the same species growing in the Colorado Rocky Mountains may attain freezing resistance to −80° C. Red maple collected during midwinter in Mississippi are reportedly resistant only to −30° C, whereas collections from Minnesota have been reported to show no injury at temperatures down to −54° C. And similar adaptation has been seen along elevational gradients. For example, Saghalien

Table 4. Tree species that show adjustment in freezing tolerance matching geographical variation in minimum temperature

Species	*Region*	*Killing temperature (°C)*
Western hemlock, *Tsuga heterophylla*	Coastal Oregon	−20[a]
	Idaho	−35 to −40[a]
Douglas fir, *Pseudotsuga menziesii*	Coastal Oregon	−20[a]
	Colorado Rocky Mts.	−50 to −80[a]
American sycamore, *Platanus occidentalis*	Mississippi	−20 to −25[a]
	Minnesota	−40[b]
Red maple, *Acer rubrum*	Mississippi	−25 to −30[a]
	Minnesota	−54[b]
White ash, *Fraxinus americana*	Mississippi	−30[c]
	New Brunswick	−43[c]
Green ash, *Fraxinus pennsylvanica*	Mississippi	−30 to −40[a]
	Minnesota	−54[b]
Eastern white pine, *Pinus strobus*	Tennessee	−39[d]
	Minnesota	−89[d]

[a]Sakai and Weiser, "Freezing Resistance of Trees," 118–26.
[b]George et al., "Low Temperature Exotherms," 519–22.
[c]N. L. Alexander, H. L. Flint, and P. A. Hammer, "Variations in Cold-Hardiness of *Fraxinus americana* Stem Tissue according to Geographic Origin," *Ecology* 65 (1984): 1087–92.
[d]D. M. Maronek and H. L. Flint, "Cold Hardiness of Needles of *Pinus strobus* L. as a Function of Geographic Source," *Forest Science* 20 (1974): 135–41.

fir (*Abies sachalinensis*) in the central mountains of Hokkaido, Japan, show a steady increase in freezing resistance with increasing elevation.[23] It is likely that many more species will join the list in table 4 as additional data become available.

WEATHERING THE WINTER DROUGHT

The desiccation problem

One of the most widely revered doctrines of winter ecology, particularly with reference to damage of trees at timberline, is that prevalent "dry winter winds" promote severe desiccation of plants exposed above the snowpack. At the turn of this century European ecologists were arguing that "cold winds possess extraordinarily strong drying power" which alone determined the limits of tree growth at both northern and high mountain timberline.[24] It is still thought that evergreen conifers exposed at timberline are universally subjected to damaging water loss during winter,[25] but such generalizations warrant some caution. While winter desiccation is an important stress in some locations, the conditions under which it occurs are often misunderstood. This is a case where intuition betrays us as we attempt to extrapolate human experience to the plant situation. The fact is that exposed plants face their greatest water loss problems under bright sunshine and calm conditions, the kind of winter day we perceive as most favorable. An increase in wind speed under these conditions would actually result in decreased transpiration. The logic of this often surprising conclusion is quite straightforward, but does require a careful examination of the factors involved in plant water loss.

Implicit in all arguments supporting desiccation as a major winter stress are a number of underlying assumptions. It is generally thought that even though leaf stomates, those tiny pores through which atmospheric gases are exchanged, remain closed throughout most of the winter, diffusion of water vapor through the protective cuticle of the leaf is still significant and is accelerated by low atmospheric humidity and high winds. It is also assumed that replenishment of water lost during winter is not possible, either because absorption is resticted by cold or frozen soils or because water movement is blocked by frozen stems or disrupted water columns. We will begin our consideration of winter desiccation by examining these assumptions, analyzing first the loss side of the water budget and then considering restrictions on the supply side.

The rate of water loss from a plant leaf, summer or winter, is directly proportional to the difference in water vapor concentration between the internal atmosphere of the leaf and the outside air, and is inversely proportional to the vapor transfer resistances offered by

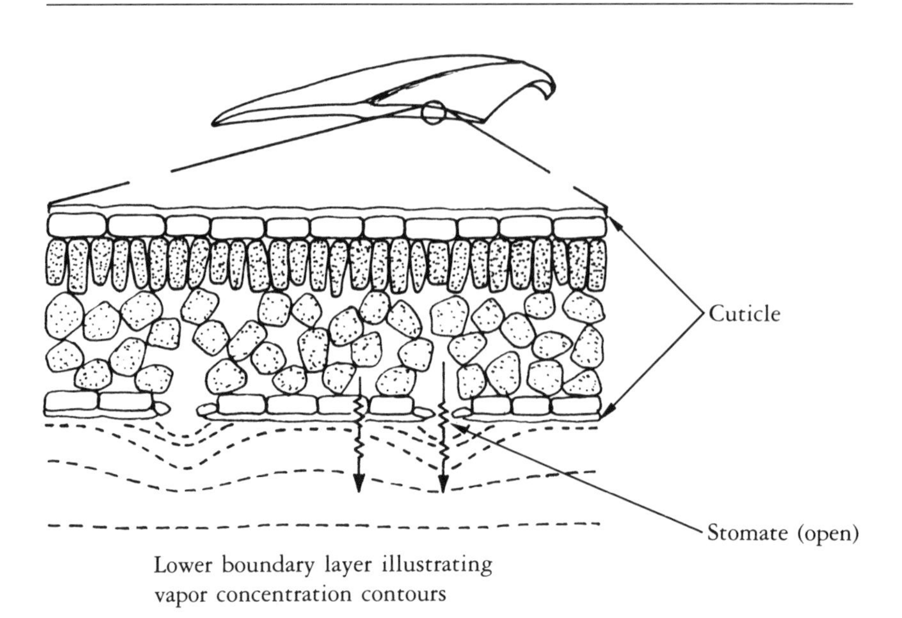

Fig. 20. A generalized model showing the pathways and resistances to diffusion of water vapor out of a leaf. The stomates and cuticle represent parallel pathways, with most of the water diffusing along the path of least resistance, through the open stomates. When stomates are closed during winter, water loss is restricted to diffusion through the cuticle, and leaf resistance thus becomes very high. Boundary layer resistance, which is provided by the thin shell of air surrounding the leaf, is influenced by leaf size and shape and by wind speed. It is essentially the same during summer and winter.

the leaf and the thin shell of air, the so-called boundary layer, surrounding the leaf. In other words, transpiration increases as the difference in vapor concentration between leaf and air increases, but decreases as diffusion resistances become greater. These pathways and resistances to vapor diffusion are illustrated schematically in figure 20.

The vapor concentration gradient that drives the transpiration process is strongly influenced by temperature differences between the leaf and outside air. Intercellular air spaces of the leaf are maintained at or very near 100% relative humidity as water evaporates from cell walls forming the terminus of water columns in the plant. The actual quantity of water vapor held in that internal atmosphere increases as temperature goes up. Therefore, the greater the elevation of leaf temperature above air temperature, as occurs when the leaf is absorbing direct or reflected sunlight, the greater the vapor concentration difference between the leaf and air. It is not at all unusual during periods of calm to find conifer needles heated several degrees above air temperature. In the extreme, needle temperatures 20° above ambient have been reported during the winter.[26] This is especially likely to occur during late spring when the sun is higher in the sky and there is still a snow-cover to reflect additional radiation to the exposed needle. The result might be a fivefold or greater increase in the driving force for water loss.

Counteracting this force for the outward movement of water vapor are the leaf resistances to diffusion provided by the stomates and the waxy cuticle covering the leaf. These resistances are easily measured and their units of "seconds per centimeter" (s/cm) are simply the reciprocal of the units for conductance (cm/s). The magnitude of leaf resistances to vapor diffusion are species dependent. The minimum stomatal resistance (i.e., when stomates are open) of many broadleaf species is less than 2 s/cm, whereas the stomatal resistance of conifers is considerably higher, usually near 20 s/cm. In contrast, the waxy cuticule that covers the outer layer of cells (the epidermis) is highly impervious to water. Hence, cuticular resistance to water vapor diffusion is several times greater in magnitude, generally ranging from 200 to 400 s/cm for conifers and up to 1000 s/cm for some desert succulents (see P.S. Nobel, Biophysical Plant Physiology and Ecology, for a summary of leaf resistance values).[27] Cuticular resistance has also been found to increase sharply with decreasing temperature.[28] As long as the cuticle remains intact, then, this protective layer is very effective at minimizing water loss when leaf stomates are closed, as is the case for evergreens throughout much of the winter.

An additional resistance to vapor diffusion is provided by a thin shell of air, the boundary layer, that is in contact with and influenced by the leaf. The thinner this surrounding layer, the more rapid the heat or vapor transfer through this zone because heat and vapor concentration gradients between the leaf and outside air will be steeper. This is analogous to the shell of air entrapped by the fur or feathers of an animal or by our own clothing, which also acts as a heat and vapor transfer zone. A leaf that is covered by dense hairs or has rolled leaf margins will likewise have a thicker boundary layer. The resistance this layer offers to vapor diffusion can be measured, and for the single "naked" leaf this resistance is generally less than 1 s/cm, although this will vary directly with leaf size and shape and inversely with wind speed.[29]

Wind currents are important because they reduce this boundary layer resistance. The effect of wind is twofold: it removes the more humid shell of air from around the leaf, thereby increasing the rate of transpiration; and it cools the leaf, thus tending to maintain temperature equilibrium between the leaf and air. In the latter case, the vapor concentration gradient between the leaf and air is reduced and consequently transpiration is decreased. The relative importance of these two opposing effects depends both on other microenvironmental factors and on the physiological behavior of the plant.

During the summer growth period, the most significant leaf resistance to loss of water under nonstress conditions is offered by the open stomates. Vapor diffusion through the cuticle is negligible as long as the stomates remain open. At this time of year the boundary layer resistance (~1 s/cm) is closer in magnitude to the leaf resistance (~2 s/cm) and any reduction of boundary layer thickness by air turbulence becomes significant in terms of increasing transpiration. In this case the effect of wind on leaf temperature, is less important.

In the wintertime, however, the relative importance of the diffusive resistances is changed significantly. Stomatal opening is not known to occur during the winter, as apparently it is prevented by low temperatures.[30] Even midwinter thaws do not appear to induce stomatal movement. With stomates closed, vapor diffusion occurs primarily through the cuticle, and the leaf resistance thus becomes

WIND AND THE WINTER-EXPOSED PLANT

To understand how atmospheric properties such as humidity and wind speed interact with the plant leaf to influence water loss during winter, it is useful to consider the transpiration process in simple mathematical terms. This has the advantage of allowing us to experiment with real or hypothetical numbers and to predict the effect of changing plant or environmental variables on water loss. In our model, transpiration is described by the simple equation

$$E = \frac{(c_l - c_a)}{r_l + r_a},$$

where E is the transpiration rate, $c_l - c_a$ is the water vapor concentration difference between the intercellular spaces of the leaf (c_l) and the air outside the leaf boundary layer (c_a), and r_l and r_a are respectively the leaf and boundary layer resistances to the diffusion of water vapor.

The driving force for transpiration is the gradient $c_l - c_a$, since water will tend to move from areas of high concentration to areas of low concentration. This gradient is strongly influenced by temperature differences between the leaf and outside air and, thus, is greatly affected by the heating of the leaf as it absorbs solar radiation. The influence of this heating on the vapor concentration gradient can be readily appreciated by consulting a table of saturation vapor concentrations for different temperatures. With the air at $-10°C$ and 50% relative humidity, and with a leaf at the same temperature as the air, the difference in vapor concentration between the inside and outside of the leaf would be 1.2 $\mu g/cm^3$. If the leaf were heated to 10° C, however, the difference in vapor concentration would increase to 8.2 $\mu g/cm^3$, a substantial increase in the attractive force for water vapor moving out of the leaf. Under these circumstances, the dissipation of heat from the leaf by wind would substantially reduce the vapor concentration gradient between leaf and air.

Opposing this driving force for water loss are the diffusive resistances offered by the leaf and boundary layer of air surrounding the leaf. During the summer when leaf stomates are open and plant resistance to water loss is generally low (e.g., 2 s/cm), a reduction in boundary layer resistance from a maximum of 1 s/cm to near zero as wind speed increases will have a significant effect on the denominator of our equation. However, during the winter r_l is generally very high (e.g., 200 s/cm), while maximum r_a remains near 1 s/cm in still air. By substituting these numbers into our model, we can readily see that a reduction of r_a by wind in this case would have very little effect on the denominator of our equation. Instead, the dominant effect of wind during the winter, when diffusive resistances are very high, is to dissipate heat from the leaf, thereby maintaining a narrower vapor concentration difference between leaf and air. The net effect of wind under these circumstances would be to reduce, rather than increase, water loss.

very high, in most cases exceeding 200 s/cm. We can readily appreciate now that when leaf resistance to water loss is very high, any reduction of the boundary layer resistance to vapor diffusion by high winds is negligible by comparison, and unimportant in terms of increasing vapor transfer (see box on "Wind and the Winter-Exposed Plant"). So as long as the leaf cuticle remains intact, the dominant effect of wind in the wintertime is to force heat dissipation. This has two important consequences: maintenance of leaf temperatures below the freezing point of cell water when air temperatures are very low; and reduction of temperature differences between the leaf and air (fig. 21*a*), with consequent reduction of the vapor concentration gradient between leaf and air. In the first case, water loss is limited to sublimation from frozen tissues, requiring greater energy input than for the evaporation of free water. In both cases, the net effect of wind is to reduce, rather than increase water loss (fig. 21*b*).

The best test of these conclusions is, of course, to follow the water balance of evergreens under field conditions where their foliage remains exposed to high winds throughout the winter. This was done for the duration of two winters at timberline on Mt. Washington[31] and for a similar period near the summit of Mt. Monadnock[32] in the northern Appalachian Mountains. Neither site produced any evidence of damaging winter desiccation in healthy foliage. In fact, on Mt. Washington, the relative water content of the leafless birch stems (*Betula papyrifera*) dropped to lower levels throughout the winter than did the foliage and shoots of exposed spruce and fir (fig. 22). Furthermore, the balsam fir at timberline maintained the same foliar water content as fir growing in a Christmas tree plantation at a sheltered location near sea level.[33]

The northern Appalachian mountains are characterized by prevalent high winds and frequent cloud cover during winter, conditions which by our previous arguments would not be expected to induce significant water loss in exposed foliage. This is not so often the case in other parts of the world, where winter desiccation is reported to be a common problem. Many instances of winter drought and desiccation damage in timberline areas of Japan, central Europe, and the western United States are the result of exposed plant parts heating above freezing due to high direct and reflected radiation loads during periods of prolonged calm. Water loss under these conditions is also often enhanced by incomplete cuticle development during the shortened growing season at high elevations, and may be further compounded by abrasion of the leaf cuticle by wind-carried ice particles. High winds have other important effects on the winter-exposed plant that will be discussed later.

*Fig. 21. The relation of water loss to wind speed in dormant leaves. Maximum water loss from winter-exposed foliage occurs under calm conditions when leaves are heated strongly by the absorption of solar radiation. If leaf resistance to vapor diffusion is very high, as with an undamaged conifer needle in winter, the effect of increasing wind speed on the reduction of boundary layer resistance becomes negligible. However, an increase in wind speed does reduce leaf-to-air temperature differences (*a*), thus reducing the vapor concentration gradient between leaf and air. The result is a decrease in water loss from exposed foliage during winter as wind speed increases (*b*). (*Source: *P. J. Marchand and B. F. Chabot, "Winter Water Relations of Tree-line Plant Species on Mt. Washington, New Hampshire,"* Arctic and Alpine Research *10 (1978): 105–16.)*

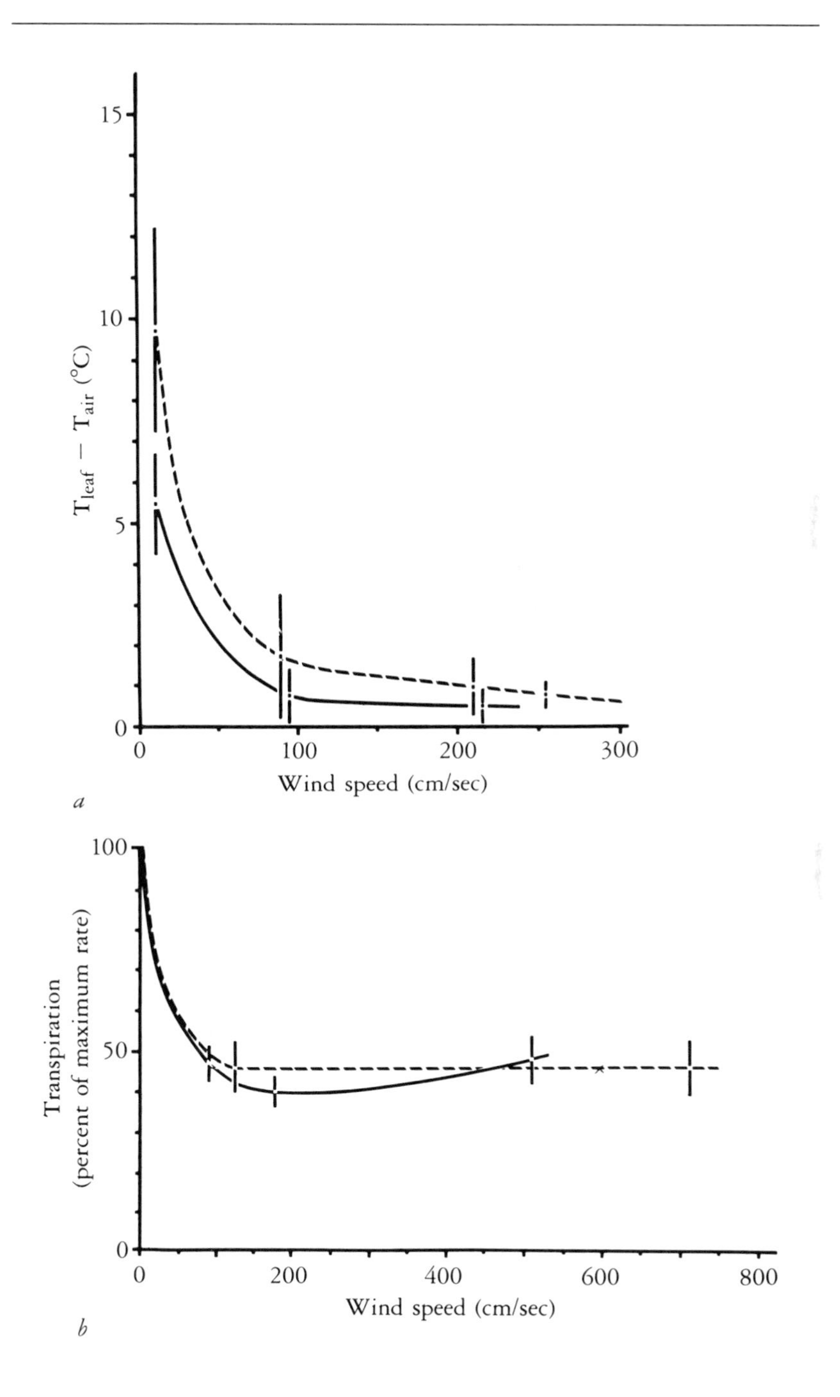
15
10
5
0
$T_{leaf} - T_{air}$ (°C)
0
100
200
300
Wind speed (cm/sec)
a
100
50
0
Transpiration
(percent of maximum rate)
0
200
400
600
800
Wind speed (cm/sec)
b

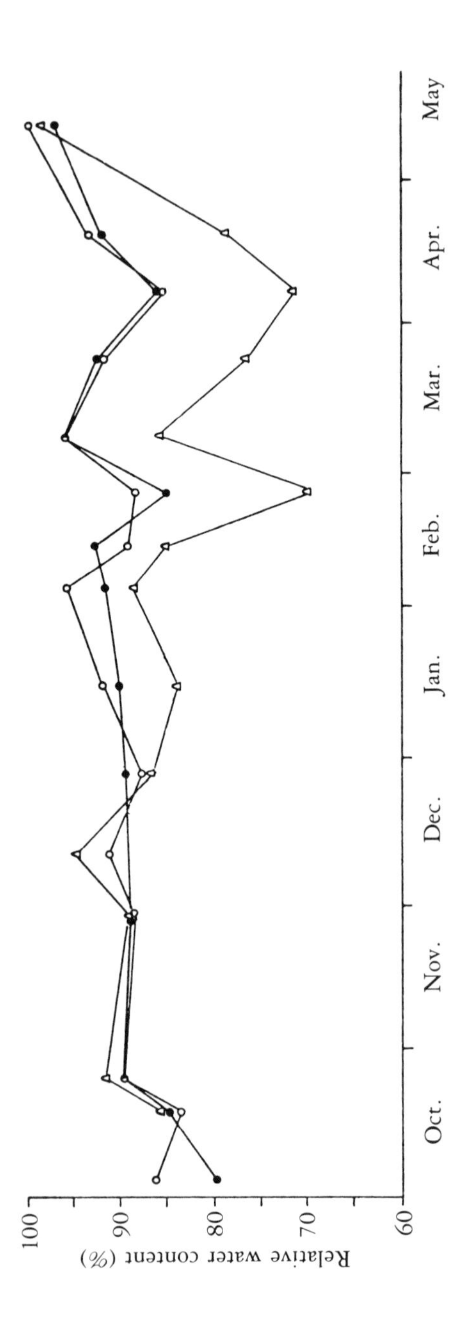
Relative water content (%)
100
90
80
70
60
Oct.
Nov.
Dec.
Jan.
Feb.
Mar.
Apr.
May

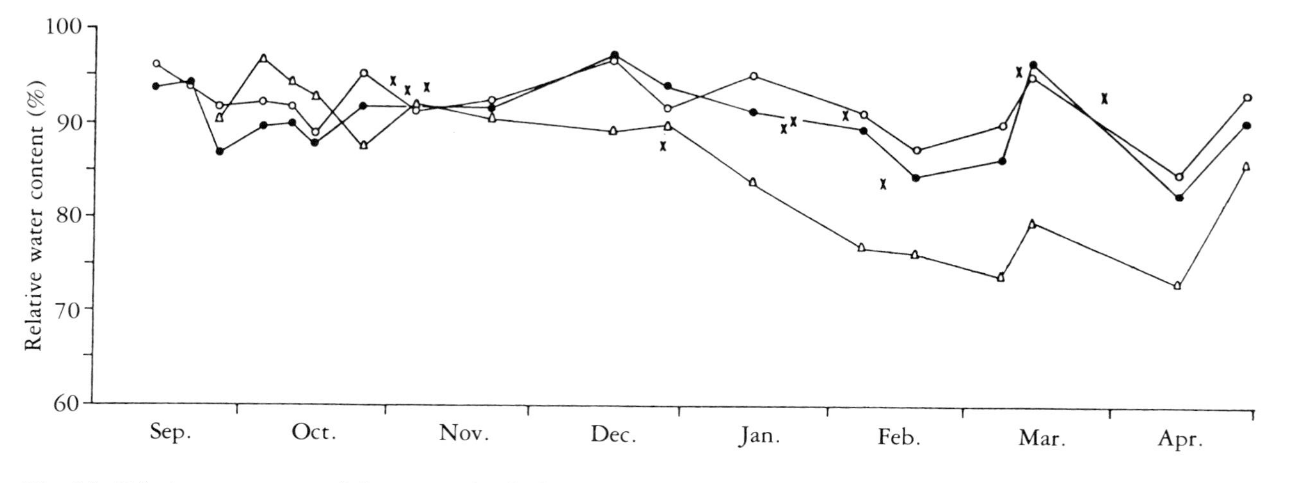

Fig. 22. Relative water content of shoots exposed to high winds throughout winter at timberline, Mt. Washington, New Hampshire. The leafless paper birch (△) exhibited the greatest seasonal decline, whereas black spruce (●) and balsam fir (○) showed the same fluctuations in water content as did balsam fir growing in a sheltered Christmas tree plantation (X) near sea level. (Source: *Marchand and Chabot, "Winter Water Relations," 105–16.)*

Water supply during winter

The indictment of "damaging water loss" implies that resupply of water during winter is not possible, a common assumption even among ecologists. However, a Norwegian scientist more than two decades ago calculated that the total water loss from a mature Norway spruce tree (*Picea abies*) during the course of one winter would exceed by nearly 10 times the maximum loss that it could survive if the foliage were not somehow replenished periodically. But the possibility of resupply is problematical. Apart from questions of where the water might come from, the likelihood of translocation by conventional pathways is limited by the supposed loss of continuity of the water column upon freezing. The problem is as follows:

Water movement in the xylem of a plant (the major conducting tissue) is a passive process that depends upon the cohesion of water in a continuous column within files of cells that serve as capillaries. Because of the high attraction of water molecules to one another the water column is effectively pulled through the xylem as a result of evaporation from various surfaces of the plant. This puts a tension on the water column much like that of a stretched rubber band. Normally, when the water in a capillary is under tension, as is almost always the case in the conducting elements of a tree or shrub, the dissolution of gases upon freezing causes cavitation—an air embolism—breaking the continuity of the water column, like cutting the rubber band. Without some means of forcing the column back together, there is no way to reestablish flow in the xylem of the plant.

How, then, might resupply of water to exposed shoots and foliage occur? H. T. Hammel provided one possible answer when he suggested a mechanism by which freezing in the xylem of conifers could happen without cavitation.[34] Water moving in the xylem of a conifer must pass through bordered pits located at the ends of each conducting cell (fig. 23). Stretching across each bordered pit is a fibrous membrane with a thickened central portion known as the "torus." Water moves freely around the torus as long as the torus remains centered in the pit. However, if anything causes unequal tension or

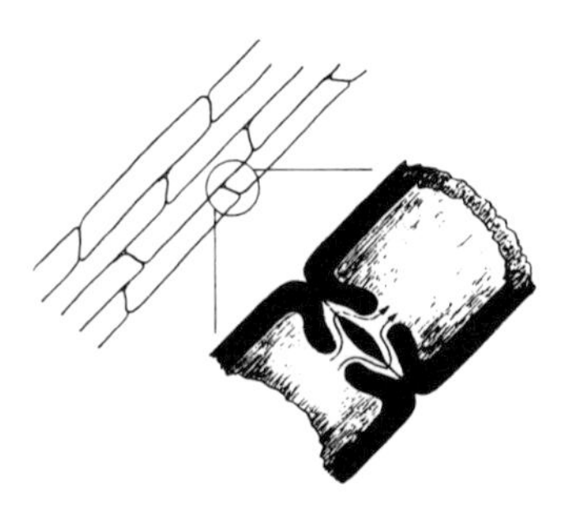

Fig. 23. Diagrammatic illustration of a pair of bordered pits between two conducting cells (tracheids) in a conifer xylem. The pits are separated by a fibrous pit membrane with a thickened central portion called the "torus." The torus acts as a valve, closing tightly against one side or the other of the bordered pit in response to small pressure changes in the xylem.

pressure distribution on either side of the torus, it deflects toward the side of highest tension, closing off the bordered pit. A pressure drop of as little as .03 MPa (5 psi) across the torus may be all that is needed to activate this check-valve.[35]

According to Hammel, this is just what happens when freezing occurs in the xylem of conifers. Because of the expansion of water upon conversion to ice, the freezing of only 1% of the water confined in a tracheid (the basic conducting element in the xylem) would reduce hydrostatic tension in that cell by about .2 MPa or 30 psi. Since there is likely to be only a slight tension in the water column at the time of freezing (atmospheric "pull" is normally low then), only a small fraction of the total water need freeze before all the tension is relaxed and the torus closes tightly. As the remaining water freezes and expands, hydrostatic pressure then builds inside the relatively rigid xylem cell, possibly reaching as high as 6 MPa, or more than 900 psi. The resistance of the cell wall to expansion means that the air bubbles now frozen out of solution will be under tremendous pressure as soon as thawing occurs, forcing them to redissolve almost immediately upon warming.

With thawing, pressure changes in the cell follow the reverse pattern. As the ice melts the pressure is gradually reduced to zero and then, with the final fraction of ice melting, the original tension gradient is reestablished and the torus relaxes to its original position. The continuity of the water column has been restored.

The key to this mechanism for preventing cavitation is confinement of water within a closed conducting element so that sufficient pressure will build with ice expansion to force resolution of dissolved gases after thawing. However, in the broadleaved trees and shrubs, the architecture of the xylem is quite different and no means of confinement is available. The cells that make up the bulk of the conducting tissue in these plants simply have perforated end walls, like sieves, through which water moves freely, with no anatomical feature at all analogous to the torus of the conducting cells in conifers. In fact, in "ring-porous" species (those that produce very large diameter conducting cells in the early part of the growing season, as opposed to the uniform small diameter conducting cells of "diffuse-porous" species), the end walls of the conducting cells are completely dissolved away so that water moves in long open-ended conduits. When water freezes and expands in these cells, it simply forces the unfrozen water to flow away from the freezing zone, relaxing tension in the water column slightly, but without ever creating the positive pressure needed to force resolution of air bubbles, even after the whole column is frozen.

The emptying of vessels upon cavitation of the water column may help explain why ring-porous trees, like the oaks and hickories, are absent from boreal forest regions where winters are so long. But what about all those diffuse-porous species that are so successful in the North? They differ only in having much smaller diameter vessels; they also lack any toruslike mechanism for isolating water in separate xylem elements. Looking again at the data in figure 22, we see periodic increases in the water content of the diffuse-porous paper birch (*Betula papyrifera*) during winter, which parallels that of the two conifers. How might this be accomplished?

For some illumination on this problem, we can look at the early experiments of J. R. Havis.[36] Using a specially constructed cooling collar placed around shrub stem segments, Havis studied the movement of water in mountain laurel (*Kalmia latifolia*) and American holly (*Ilex opaca*) during freezing. By cooling a stem through the first freezing exotherm and then applying acid fuchsin dye solution to the cut end of the stem, he was able to demonstrate water movement through the wood even though the xylem was blocked by ice. He was also able to show quantitatively that the rate of water movement did not slow until the stem was cooled through a second freezing plateau, at which time all flow then stopped. Havis interpreted this to mean that the primary pathway of water movement in the stems of these dormant plants was through the interconnected cell walls rather than through the conducting vessels themselves. If that is the case, this may circumvent the problem of cavitation. In this way, plants maintain some supply of water during winter whenever temperature conditions allow, until the water column can be restored through renewed cambial divisions in the spring.

Havis was not the first to demonstrate dye movement in stems under subfreezing temperatures. G. Hygen in Norway used a similar tracer under field conditions to make a case for resupply of water to drying foliage of Norway spruce.[37] Injecting dye into a tree just above the snow surface, he found that water moved upward during mid winter at a rate of 3% of the summer flow. Thirty years earlier, a fellow Norwegian, N. Polunin had also reported that dye moved through roots of silver birch (*Betula adorata*) imbedded in solidly frozen soil.[38]

With the increased use of radioactive tracers and heat pulse techniques in the late 1960s, several more attempts were made to measure water uptake and conduction in the field during winter, usually with the same result. Water movement was noted whenever stem temperatures were above freezing, even though soils were still frozen. In one of the more detailed studies of its kind, P. Owston and his colleagues followed water uptake seasonally in lodgepole pine (*Pinus contorta*) and red fir (*Abies magnifica*) by introducing radioactive phosphorus (^{32}P) to the clipped ends of lateral roots and then monitoring the ascent of the tracer with a portable scintillation counter.[39] During winter they found movement in only two of five red fir and none in the five lodgepole pines injected. In this case they attributed the lack of water conduction to the freezing of exposed lateral roots. In April, which was the coldest month of their study period, they found only one radioactive path, again in red fir, for 10 new injections made at the beginning of the month. However, the tracer that was injected in April began moving in all trees in early May, even though the snow was still 3 m deep and the soil temperature remained around 1° C. This suggests that it was low temperatures in the bole above the snowpack rather than low soil temperatures that was restricting water movement.

Whether or not water absorption from frozen soils is possible remains problematical. The presence of dissolved solutes may lower the freezing point of soil water a fraction of a degree;[40] and for the same reason that water in a plant may supercool, some fraction of the soil water may also remain in liquid form below the freezing point. It is possible that retention of water on soil particles with fairly high tensions may preserve a liquid film around the particle, while free water in the soil pore spaces is frozen. As long as the tension in the water column of the plant root is greater than that in the soil, then this unfrozen water should be available to the plant. Even when soils are not frozen, however, the mobility of soil water during winter is restricted by low temperatures. The viscosity of water at 0° C is twice as great as it is at 20° C. Nevertheless, the slow migration of water toward plant roots may be sufficient to keep pace with foliar losses at this time of year.

The fact that conduction may resume whenever aboveground tissues thaw suggests another source of water that may be available to drying foliage. That is the water stored in the plant itself; water held in conducting cells, the cell walls, and intercellular spaces of the tree or shrub. It has been estimated that the sapwood of a tree, generally representing 20 to 40% of the radial dimension of the tree, holds enough water to satisfy all its requirements at a moderate transpiration rate for one week.[41] At lower transpiration rates the supply would, of course, last longer. With cuticular water loss from dormant conifer foliage being some 40 times less than that of actively transpiring foliage, water stored in the sapwood might well be sufficient to carry the tree through the whole winter.[42] This assumes that opportunity for mobilization of water occurs with sufficient frequency throughout the winter. Given that conifer needles are often observed to be above freezing even when ambient temperatures are lower, it seems reasonable to expect that exposed parts of the bole and branches would likewise be warmed. It is likely, then, that the upward movement of water reported in some conifers during winter represents internal redistribution rather than actual water loss from foliage at the time of measurement.

One other possibility remains for replenishment of water, and that is through foliar absorption of water vapor. There is little question that plants that remain buried in the snowpack all winter gradually make up any water deficit with which they may have entered the season. As long as the snowpack air spaces are saturated with water vapor and the plant beneath the snow remains at the same temperature as the snowpack (i.e., does not heat by absorption of light penetrating the snow), then the direction of water vapor transfer will always be from the air to the leaf. However, for the tree or shrub exposed above the snowpack, opportunity for foliar absorption may be very limited, even where branches are laden with snow. Even a leaf under some water stress will have an internal atmosphere of greater than 99% relative humidity, and it is not often that ambient humidity will exceed this. Nevertheless, there remains the possibility that as long as some portion of the plant's foliage lies buried under the snowpack, enough absorption may occur to benefit exposed foliage through internal redistribution of water.

Recently, my students and I attempted an experiment to determine the relative contribution of these various pathways to resupply. While not conclusive yet, the results to date have been interesting and provide some insight into this problem. In our study, three treatments were designed to eliminate systematically one pathway of water absorption or another in red spruce (*Picea rubens*) and balsam fir (*Abies balsamea*) trees, each about 3 m in height. The treatments, described below, were initiated in January after a snowcover of approximately 1 m had accumulated, burying a substantial portion of the foliage of these trees.

In the first treatment we carefully cut trees at their base with a long-handled pruning saw while the snowpack was left as undisturbed as possible. The base of the tree was sealed by inserting a jar cap filled with melted parafin between the cut surfaces. The trees were then guyed with rope while the lower branches remained buried in the snow. The intent of this treatment was to elimate root absorption while still allowing for the possibility of foliar absorption and internal redistribution of water.

In the second treatment, the trees were left intact to allow for root uptake, but all branches below the snow were pruned to eliminate any foliar absorption. This resulted in the removal of about one-third of the foliage of each tree, but did not alter the total amount of foliage exposed to water loss above the snowpack.

In the third treatment, we made no alterations to the tree itself, but we removed all snow from around the tree and kept the ground clear for the remainder of the winter. This resulted in freezing the soil to considerable depth. The intent here was to minimize root uptake and translocation while at the same time eliminating any possibility of foliar absorption.

For the duration of the winter we monitored the water content of exposed foliage on all trees, including undisturbed control trees, weekly. Each winter that we conducted this experiment the results were the same, typified by the data shown in figure 24. Throughout the winter no treatment produced any substantial difference in foliar water content. Exposed foliage of the experimental trees periodically gained or lost water in the same manner as the control trees—until very late spring. As air temperature warmed and snowmelt accelerated, all treatments except one showed a pronounced increase in water content. At this point the basally cut trees experienced a dramatic decrease in foliar water content.

Until late spring the common denominator which kept the foliage of all trees fully hydrated was apparently stored reserves. Regardless of treatment, all trees appeared to have sufficient water available internally to carry them through the winter. Only when internal water was exhausted and the basally cut trees had no means of uptake did the difference due to that treatment appear. While

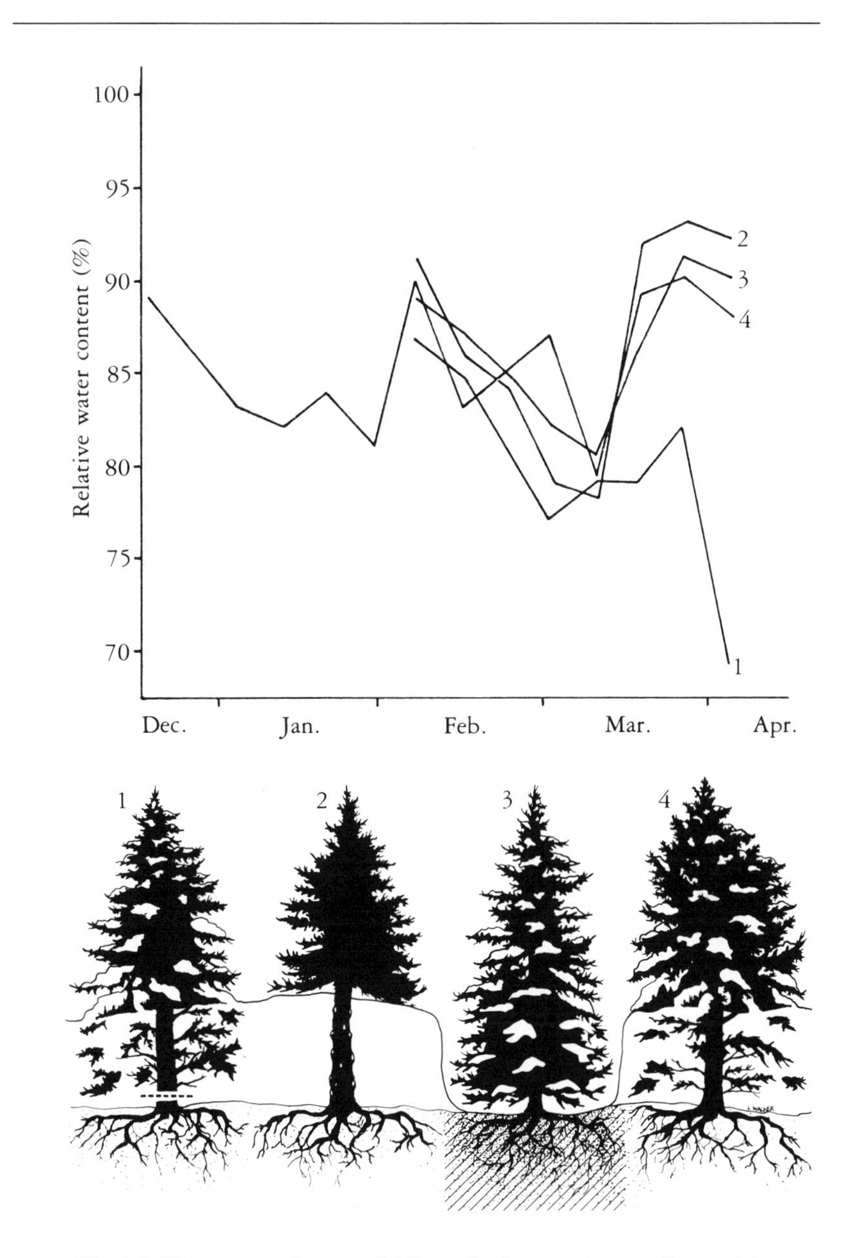

Fig. 24. Water content in exposed foliage of red spruce, Picea rubens, *following the treatments indicated.*

treatments designed to show foliar absorption or root uptake from frozen soil throughout the winter did not result in increased foliar water content, it is clear that resupply of water to these other trees was possible sometime before the basally cut trees exhausted their internal reserves.

It should be reiterated at this point that a plant frequently exposed to high direct and reflected solar radiation during periods of prolonged calm may have a problem maintaining a favorable water balance during a long winter. Under these circumstances any opportunity to absorb water, whether by roots or foliage, may be advantageous. It is clear that in some situations, most notably at timberline in the European Alps and in some mountain ranges of the western United States where stunted trees may have limited internal reserves, severe desiccation may be an important factor in the overwintering success of plants.

THE EVERGREEN ADVANTAGE

We have seen that retention of foliage through the winter may prove, under certain conditions, to be something of a liability. If this is the case, then why be evergreen?

It is often supposed that the advantage of the evergreen habit in conifers is to enable continuance of photosynthesis throughout the winter. Since chlorophyll in these trees does not undergo the seasonal degradation so obvious in the autumn leaves of deciduous trees, it is presumed to remain functional in its light-trapping capacity—and green is equated with photosynthesis. However, it is probably safe to say that midwinter photosynthesis in northern conifers is more the exception than the rule, although there is little doubt that the photosynthetic season may be extended considerably by the evergreen habit.

Whether or not photosynthesis occurs at all in exposed conifer needles during the winter seems to depend as much on external factors as on any internal plant cycles. In milder winter climates a number of conifers have shown small positive net photosynthesis whenever temperatures climb above freezing. Austrian pine and Norway spruce at low elevations in central Europe have been found to photosynthesize on warm winter days at rates sometimes equalling 25% or more of their summer and fall rates. In the subalpine forests of New Zealand's Southern Alps, some evergreen trees show only a brief photosynthetic dormancy. There the occurrence of daily frosts in late autumn (June) is accompanied by a complete cessation of photosynthesis in mugo pine, mountain beech, and lodgepole pine. Barely two months later, though, while nighttime temperatures are still several degrees below freezing, this dormancy is broken. Photosynthesis measured when daytime needle temperatures approach 12° to 15° C can be as much as 10% of maximum late summer rates for mountain beech and 35% for the pines. Still higher photosynthetic rates have been recorded during late winter at lower elevations.[43]

In more severe winter climates, the situation is very different. At elevations near timberline in the European Alps, the photosynthetic machinery of Austrian pine appears completely shut down during midwinter and the trees are not at all able to take advantage of brief winter thaws.[44] Likewise, with bristlecone pine growing near timberline in the White Mountains of California, no photosynthesis occurs during winter even though daytime temperatures may rise above freezing for several days in any winter month. The carbon balance (photosynthetic uptake of CO_2 minus respiratory loss) of these trees remains continuously negative from mid November to the end of April.[45]

Indirect evidence indicates that the photosynthetic machinery of other northern conifers may also be inactivated during the winter months. In balsam fir, the chloroplasts of leaf mesophyll cells (beneath the epidermis) are clumped together during midwinter rather than being distributed evenly around the perimeter of the cell for maximum light-trapping efficiency, as they are in the summer months.[46] In black spruce, the leaf stomates through which the CO_2

for photosynthesis diffuses show sensitivity to changing light levels (itself a photosynthetic response) only after preconditioning for several days at constant above-freezing temperatures.[47]

Even where conifers are not able to take advantage of brief winter thaws, their photosynthetic machinery may be reactivated surprisingly early in the spring. Mesophyll cells of balsam fir begin to show signs of structural reorganization, including chloroplast realignment, two months before bud break. Even at timberline pronounced increases in needle starch accumulation as a result of photosynthetic activity can appear in March, long before winter shows any signs of yielding to spring. This suggests that the photosynthetic process is not coupled to any of the metabolic processes controlling acclimation and freezing resistance. It seems reasonable to conclude that evergreen conifers benefit, at the very least, from an extended photosynthetic season, if not a continous one.

What, then, might we expect with other evergreen plants, particularly those that remain under the snowpack throughout the winter? With regard to temperature, subnivean plants enjoy a much more benevolent environment—one which is also free from problems associated with water stress. We know that many plants remain or turn green and actively grow under the snow, sometimes even forming flowers before snowmelt. In many cases this is accomplished through the utilization of stored energy reserves; but is it not possible that some growth is supported by photosynthesis under the snow? Although light levels under the snowpack are much reduced, the light that penetrates to greatest depths is of a wavelength near optimum for chlorophyll absorption. Chlorophyll synthesis has been measured in plants responding to the light penetrating as much as 80 cm of snow. Furthermore, the elevated CO_2 levels sometimes found under the snowpack (a point that wil be discussed later) could reduce the light-compensation point such that positive net photosynthesis might be possible under lower light levels than normally expected.

Too few attempts at measuring photosynthesis in subnivean plants have been made to allow any generalizations yet, but the evidence points to at least limited photosynthetic activity under the snow. L. Tieszen was able to measure limited carbon fixation in three arctic herbaceous species just before snowmelt, but also found that the plant leaves lacked their full complement of enzymatic activity, as is often the case with immature leaves, and this reduced their photosynthetic potential.[48] In the laboratory, Erici Mäenpää simulated the light and temperature conditions that prevail under 30 cm or more of snow during the springtime in northern Finland and succeeded in measuring positive net photosynthesis in the green stems of bilberry (*Vaccinium myrtillus*) under those conditions.[49] And in Austria, W. Tranquillini reported "break-even" CO_2 assimilation in Austrian pine seedlings under 50 cm of snow.[50] It is clear from these examples that some, perhaps even the majority, of the subnivean plants that remain green through the winter have the ability to utilize light energy penetrating the snow.

Plants that retain their green leaves during the winter may not be the only ones to benefit from an extended photosynthetic season. More than 50 tree species, the majority of them deciduous, are known to have chlorophyll in their bark tissues.[51] In aspen, the most studied of these species, the bark may comprise up to 15% of the total photosynthetic surface area and may contribute substantially to the annual carbohydrate gain of the tree. Photosynthesis in aspen bark has been detected in winter, and it is speculated that this may enable this species to compete with the northern conifers during the long leafless season. In fact, winter photosynthesis in bark tissues may be the rule rather than the exception. T. Perry has measured photosynthesis in leafless twigs of sycamore, oak, and pecan immediately after thawing.[52] Similarly, Mäenpää in Finland has found evidence of photosynthesis in the green stems of bilberry under subnivean conditions, as noted earlier. A substantial amount of chlorophyll has also been found in the bark tissues of larch, and it is interesting to speculate as to whether or not winter photosynthesis might compensate for the deciduous habit of this conifer as well.

It seems reasonable to assume that winter photosynthesis in bark tissues operates under the same constraints as in evergreen leaves and thus would vary with the length and severity of the winter season. There is, however, one important difference: the resistances to diffusion of CO_2 that limit photosynthesis in leaves would not be the same in bark tissues. In fact, much of the CO_2 supply to the bark chloroplasts may come from internal sources. It has been estimated, for example, that in aspen 50 to 75% of the CO_2 released by respiration is re-utilized in bark photosynthesis.[53] If this is typical of other species as well, then it could make a significant difference in the amount of winter photosynthesis taking place.

Whether or not the products of photosynthesis in bark tissues are transported to any other parts of the tree remains a question because of the separation of these outer layers from the phloem tissue, which is the major transport system for nutrients in the woody plant. It may be that carbohydrates produced by the bark choloroplasts are utilized primarily in the outer layers of the plant. Nevertheless, any opportunity for photosynthesis, however limited, would help offset the continuous respiratory loss of carbohydrates in woody plants during the winter and would thus be of some adaptive value in northern areas.

MECHANICAL PROBLEMS: THE BRUTE FORCES OF WINTER

The mechanical aspects of the winter environment—heavy snow loads, blowing ice particles, even the browsing activity of mammals—pose a number of problems for plants. The effects are often direct and readily observable, but at times may be indirect and subtle, becoming apparent only sometime after the damage has occurred. In some places these forces have a profound influence on the distribution of plant communities and the development of plant growth form.

Snow loading is always a potential hazard to trees and shrubs, especially in boreal forest regions where winters are characterized by periods of prolonged calm and snow remains on branches for an

Fig. 25. Burdened by winter. Snow loading poses a particularly severe problem to conifers in regions where snows are heavy and where periods of prolonged calm result in long retention by the tree. In northern Finland, where winter storms are fed by moisture from the Baltic Sea, tree breakage under extreme snow loading, as shown here, is considered to be the major limiting factor at treeline. (Photo by Kari Laine.)

extended time. In northern oceanic climates where milder winter temperatures and high moisture result in more frequent ice storms and heavy falls of wet snow, repeated bending and breaking of trees is common (fig. 25). The pronounced spire form of conifers in northern regions helps minimize snow loading, but the weight of snow retained in some cases is still impressive. In the spruce uplands of Finland, where winter precipitation is influenced by the Baltic Sea, a tree 12 m in height can accumulate a mass of snow and ice nearly 50 cm thick and weighing 3000 kg! So dominant a force is this in the hill country of southern Lapland that tree breakage there is considered to be the major factor limiting forest stand development at timberline.[54]

In areas continually exposed to high winds, it is not heavy snow loads, but rather the abrasive force of blowing ice particles that shapes the forest stand. Just above the snow surface, the stinging force of ice particles whipped by the wind is particularly strong, and its effect is obvious wherever objects protrude above the snowpack. A wooden post covered with several layers of paint and protected on top with an inverted can provides graphic evidence of the abrasive force of this ice blast (fig. 26*a*). Likewise, bark on the windward side of exposed trees is deeply pitted, even completely worn away (fig. 26*b*). Foliage is stripped, sometimes cleanly and sometimes only partially, leaving many broken needle stubs (fig. 26*c*). In some cases ice blast removes foliage already damaged by desiccation or freezing injury; but in more severe situations healthy foliage and bark are abraded away by the force of wind-carried ice particles.

With damage to fresh tissue, water loss may then be accelerated, further aggravating the mechanical injury.[55] The initiation of wound-healing processes, including callous tissue formation, is apparently delayed during winter, resulting in reduced water content of affected branches due to the exposure of open-ended vascular tissue (fig. 27). With continued drying, the loss of foliage may be compounded. If this desiccation is not lethal, the branch may rehydrate the following spring. The reduction of photosynthetic capacity accompanying such injury, though, may be an important factor hindering the recovery of conifers because much of the shoot growth during the early part of summer is dependent upon the starch reserves accumulated in the older foliage during late winter and spring.

Fig. 26. The abrasive force of blowing ice. On a windswept slope in the Southern Alps of New Zealand, blowing ice particles abrade through several layers of paint into the wood of this post where it is not protected by the inverted can (a). *The same abrasive force dramatically sculpts the boles of trees exposed to years of ice blast in the Rocky Mountains* (b) *and strips healthy foliage from exposed conifers at timberline in the White Mountains of New Hampshire* (c). *(Photo credits:* 26a *and* 26c *by Peter Marchand;* 26b *by Friedrich-Karl Holtmeier.)*

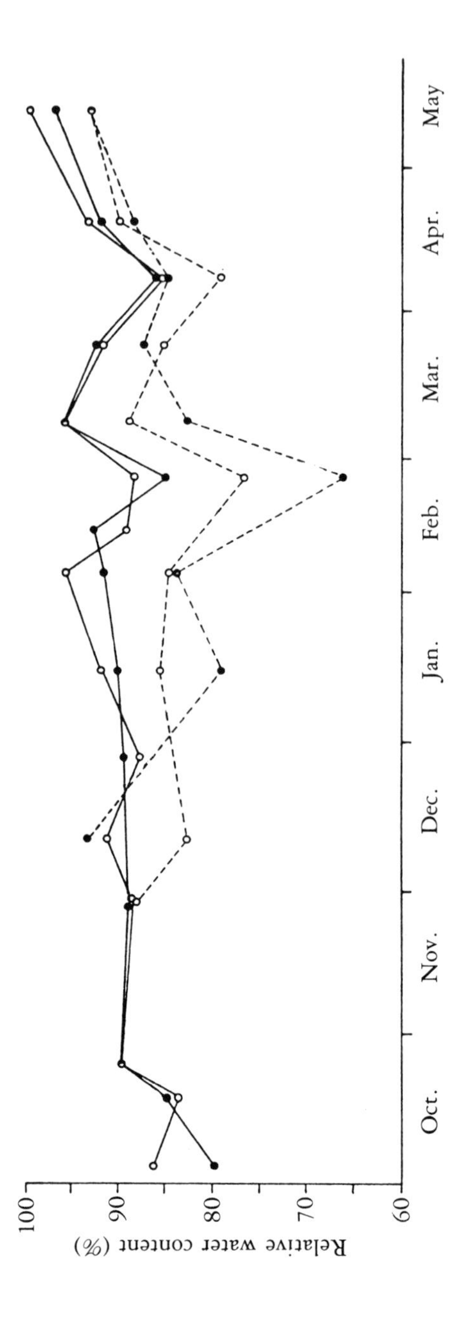
Relative water content (%)
100
90
80
70
60
Oct.
Nov.
Dec.
Jan.
Feb.
Mar.
Apr.
May

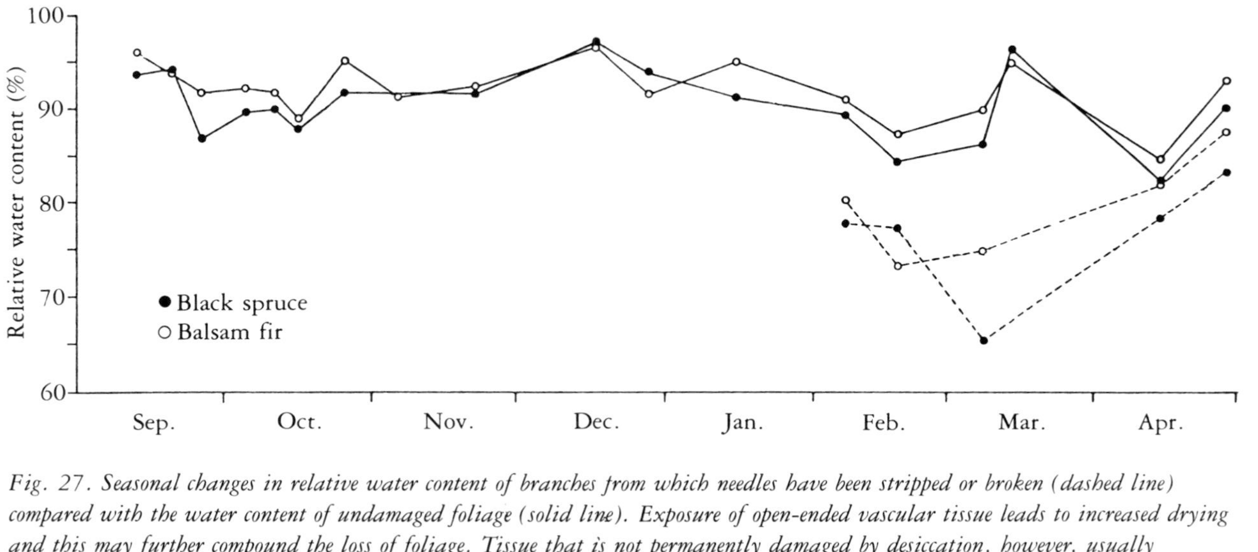

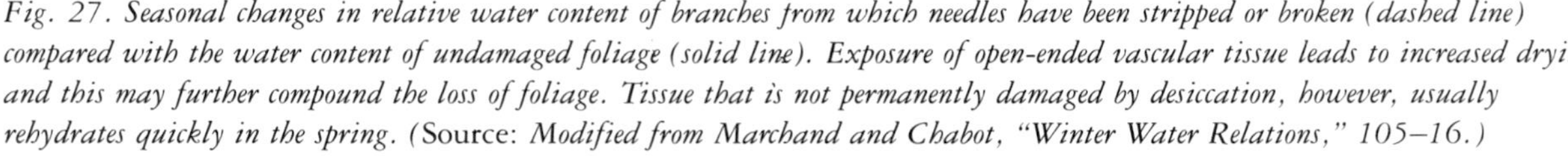

Fig. 27. Seasonal changes in relative water content of branches from which needles have been stripped or broken (dashed line) compared with the water content of undamaged foliage (solid line). Exposure of open-ended vascular tissue leads to increased drying and this may further compound the loss of foliage. Tissue that is not permanently damaged by desiccation, however, usually rehydrates quickly in the spring. (Source: *Modified from Marchand and Chabot, "Winter Water Relations," 105–16.*)

a

b

Fig. 28. Tree forms shaped by snow and wind. Mutual protection and increased snow packing in the downwind direction result in the growth of wedge-shaped islands or "hedges" (a) *on exposed slopes in the Rocky Mountains. "Table trees," this one in Alaska* (b), *form where lower branches remain protected under the snowpack, but wind-carried ice particles completely remove foliage just above the surface of the snow. When trees grow through this zone of maximum abrasion,*

c

d

they sometimes take on a "mop head" form, a process referred to as "broomsticking," exhibited by trees on this wind-exposed ridge of Mt. Washington, New Hampshire (c). *Some species, such as black spruce* (Picea mariana), *have sufficient genetic flexibility that they may grow completely prostrate under extreme environmental pressures* (d), *thus finding complete protection even under very shallow snowcover. (Photos by Peter Marchand.)*

The outward effect of this abrasive damage is the development of several distinct growth forms, usually prominent near timberline where trees are most exposed. Wedge-shaped islands or "hedges" form where the windward edge of a group of trees is severely damaged by either extreme desiccation from exposure to strong sunlight or by ice blast, but where mutual protection of branches and greater snow packing to the leeward side result in increased growth in the downwind direction (fig. 28*a*). "Table trees" occur when foliage survives beneath the snow but is almost completely abraded away at the snow line. Where some protection is afforded to the terminal bud by a surrounding cluster of lateral buds, vertical growth may persist until the tree eventually reaches above the zone of abrasion and retains a healthy top (fig. 28*b*). "Broomsticking" is the term sometimes used to describe this process that results in a branchless, windswept trunk with only a "mop head" cluster of foliage at the top and a less pronounced "table" of foliage at the base. (fig. 28*c*). Mat forms develop where species with sufficient genetic flexibility, such as black spruce, grow completely prostrate under conditions of extreme exposure, without ever producing a vertical leader, and are thus completely protected by very little snow (fig. 28*d*).

In a problem closely related to ice abrasion, the frequent deposition of rime ice at high elevations, accompanied by strong winds, also results in heavy wintertime foliage loss. Rime ice forms when supercooled cloud droplets freeze upon impact with cold surfaces as clouds sweep across mountain slopes. Resulting accumulations of ice, building on the windward sides of all exposed objects, sometimes form broad featherlike vanes of great intricacy and beauty (fig. 29*a*) that wreak havoc as winds shift and break the ice-encrusted foliage. The preponderance of green litterfall in coniferous forests under this influence usually becomes evident in the springtime as litter appears at the surface of the melting snowpack. Litter traps placed under the snowpack in subalpine balsam fir stands indicate that as much as 20% of the foliage of trees exposed at the edges of canopy gaps may come down during the winter (fig. 29*b*). For trees growing in a marginal environment and under stress from other

a *b*

*Fig. 29. Rime ice accumulation and litterfall in subalpine forests. Supercooled water droplets freeze on impact as clouds sweep across cold mountain slopes, encasing exposed trees in glistening ice (*a*). Shifting winds break the encrusted branch tips, resulting in substantial loss of green foliage during the winter (*b*), foliage which would later have supported new growth of the tree. (Photos by Peter Marchand.)*

factors, this loss may pose a significant problem and appears to be a major factor in the death of trees in subalpine fir forests of the northern Appalachians.[56]

A less destructive result of snow drifting and deep accumulation in many subalpine areas of the Rocky Mountains in the United States is the development of rather striking "ribbon forests." These forests are characterized by long bands of trees running perpendicular to the direction of the prevailing winds and separated from each other by wide corridors of moist subalpine meadow. Snow is a causal factor, but its influence in this case is on tree reproduction rather than destruction of existing stands. Trees growing in the protection of a ridge line may accumulate huge drifts of snow in its lee, snow that has blown off the exposed alpine tundra upwind. These drifts, sometimes 6 m or more in depth (fig. 30), may persist through

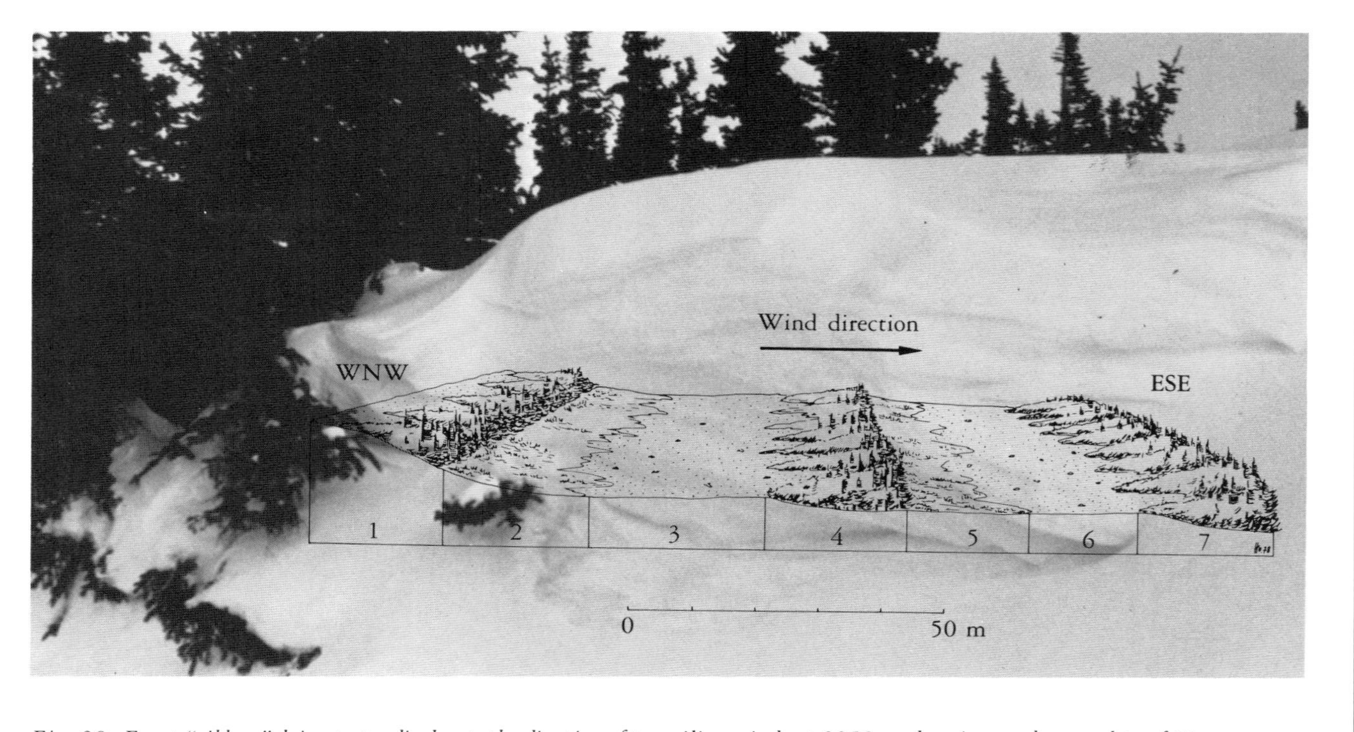

Fig. 30. Forest "ribbons" lying perpendicular to the direction of prevailing winds at 3350 m elevation on the east slope of Mt. Audubon, Rocky Mountains, Colorado. Zones 1, 4, and 7 of inset depict stands of wind-flagged Engelmann spruce and subalpine fir that accumulate huge drifts of snow in their lee (photo). These are separated by "glades" (Zones 2–3 and 5–6) in which tree seedlings are uncommon because of the very short snow-free season. (Drawing courtesy of F. K. Holtmeier. Photo by Peter Marchand.)

almost the entire following summer. Wherever the melting edge is located in late June or early July, soil temperature and moisture conditions may be optimal for the establishment of tree seedlings, which grow to form the next ribbon downwind. These seedlings will be protected under the drift again during the following and subsequent winters, until they are tall enough to stand above the snow and begin to influence the process themselves. Where the melting snow exposes open ground much later in the growing season, only the herbaceous meadow vegetation is able to establish itself, and so a snow glade forms and is maintained by the perennial drifts as long as the forest ribbon upwind remains intact. Disturbance caused by a fire or "blowout" in the natural snowfence will change the snowdrift pattern and may then result in a shift in the ribbon and glade pattern, with tree seedlings becoming established in the glade.

4 ANIMALS AND THE WINTER ENVIRONMENT

Birds and mammals that remain active throughout winter in the North face a special problem: to survive they must maintain a body temperature within quite narrow and relatively high limits. This means that the animal must produce enough heat by metabolism of its food or fat reserves to offset that which is lost to its cold surroundings. At a time when mobility is restricted and food resources are scarce, survival of the cold often becomes equated with heat conservation. To appreciate fully the problems that homeotherms face at extreme low temperatures and the energetic advantages of their many physiological and behavioral adjustments during the wintertime, we must understand the details of how heat is exchanged between the animal and its surroundings. This will require consideration of how the various modes of energy transfer—conduction, convection, radiation, and latent heat exchange—are important to the animal wintering in the North.

THE BASICS OF ENERGY EXCHANGE

The first of these energy transfer processes, heat exchange by conduction, involves the transfer of energy through molecular collisions. This requires intimate contact of the media involved in the heat transfer—for example, between skin and air or the foot surface and ground (in the case of an animal). Heat transfer by conduction is, therefore, very much dependent upon the total surface area of exposure. The rate of heat loss by conduction also depends upon the efficiency with which heat moves through the conducting medium and is influenced by the temperature difference between the heat source and heat sink as well. In general, high density materials have a high thermal conductivity, transferring heat much more rapidly than low density materials. The earlier discussion of heat transfer in a snowpack revealed that the density and thermal conductivity of a porous material like soil or snow often varies according to water content or the amount of air-filled space in the material. The same is true of an insulating layer of fur or feathers. Thus, the wetness of an animal's fur, for example, or the amount of air entrapped in the fluffed feathers of a bird will greatly affect its comfort at low temperature. Apart from material differences in efficiency of heat transfer, the temperature gradient through the transfer zone will also influence the rate of heat exchange. When two objects have greatly different temperatures, the rate of heat flow by conduction from one to the other will be much faster compared to when the two objects have nearly the same temperature.

Given these variables it is possible to estimate the amount of heat loss by conduction from an animal in a given situation. We must know only something about its total surface area of exposure, the thermal conductivity of its skin and fur or feathers (i.e., its thermal transfer layer), and the temperature of its surroundings. These variables are related mathematically by the simple equation

$$Q_c = kA\frac{(T_b - T_a)}{d},$$

where thermal conductivity (k) is measured in units of W/cm °K, surface area of exposure (A) is measured in cm^2, and the temperature gradient between the animal's core (T_b) and outside air (T_a), divided by the distance (d) in cm between the two measurements, is recorded in °C (or °K), giving us total heat loss (Q_c) in watts. We will see momentarily how helpful this model is in understanding an animal's behavior at low temperature.

The second mode of energy transfer, heat exchange by convection, involves the transfer of energy via a moving fluid, specifically air or water. In this mode, heat is gained or lost via the transfer of a parcel of air or water (that was initially heated or cooled by conduction) from one point to another. This movement may result from density differences in the fluid (free convection) or from pressure exerted on it (forced convection). As with conduction, the total area of an object exposed to the moving fluid and the temperature difference between the surface of the object and the fluid are both important in determining total heat loss. To quantify heat exchange by convection, we also need a measure of the efficiency of heat gain or loss for a given orientation in the moving fluid. This measure is often expressed as a convection coefficient and is the equivalent of the thermal conductivity of the surrounding fluid. The rate of heat transfer by convection in air is, thus, influenced by the thickness of the boundary layer enveloping the object, and this in turn is influenced by the size and shape of the object, by its surface roughness, and by the wind speed. The equation describing heat transfer by convection is similar to that for conduction, with k taking on the value of thermal conductivity of air or water and d representing the thickness of the boundary layer across which the temperature difference between surface and surroundings is measured.

The third mode of energy exchange is by radiation. Unlike conduction and convection, radiant energy transfer does not involve molecular collisions and does not require any fluid medium. Radiation is simply the propagation of energy through space, as is the case with visible light energy, and can occur in a vacuum. The two properties of any object that determine the amount of energy it emits by radiation are its temperature (independent of ambient temperature) and its characteristic efficiency of radiant energy transfer, termed its "emissivity." Any object whose temperature is above absolute zero radiates energy in direct proportion to its temperature, with most natural objects emitting radiant energy at an efficiency of between 90 and 99% of the maximum possible for their tempertures. By measuring the temperature (T) of an object and knowing or estimating its emissivity (ε), we can, with reasonable accuracy, calculate its total energy loss via radiation using the equation

$$Q_r = \varepsilon \sigma T^4,$$

where σ is the Stefan-Boltzmann constant (5.67×10^{-8} W/m^2 °K^4). This tells us that for most homeotherms heat loss by radiation remains fairly constant throughout the year since body temperature varies only slightly over time. Using the same equation, we can also easily calculate the amount of heat gained by an animal through the absorption of radiant energy emitted by its surroundings.

The final mode of energy transfer is by latent heat exchange. Latent heat is the amount of heat energy either tied up or liberated with phase changes of water. For example, the conversion of 1 g of ice at 0° C to water at the same temperature requires the addition of 335 joules of heat energy (i.e., 335 joules are tied up in the gram of water upon melting). This quantity is referred to as the latent heat of fusion. It is important to understand that this addition of energy does not alter the temperature any, but is required just to switch phase from solid to liquid. To convert 1 g of water at 20° C to vapor requires the addition of approximately 2450 joules of heat energy, the latent heat of vaporization. When the reverse process occurs, the same amount of heat is liberated. Latent heat exchanges in nature involve one of these phase changes, usually accompanied by a migration of water or vapor from source to sink.

The practical implications of latent heat transfer are many. Through this mechanism, for example, normal respiration in warm-bodied animals sets up something of a heat pump, with each exhaled breath removing heat by evaporation from the wet surfaces of the lung—heat that is robbed from the body as water changes phase from liquid to vapor. For those of us with active sweat glands in the wintertime, the benefits of reducing evaporative heat loss can be readily appreciated. The old trick of putting your feet in plastic bags before pulling on your socks and boots works well because the plastic barrier prevents the diffusion of vapor outward, thereby reducing the loss of latent heat that is tied up in vapor molecules (fig. 31). By keeping your socks dry at the same time, the rate of heat transfer by conduction is also reduced. The resulting condensation inside the bag may keep your feet wet with perspiration, but it also keeps body heat where you want it most. For the same reason, the common practice of installing vapor barriers behind the inside walls of our homes reduces latent heat loss and results in considerable fuel savings during the winter. Note that in both cases it is important that the vapor barrier be located on the warm side of the insulation. If

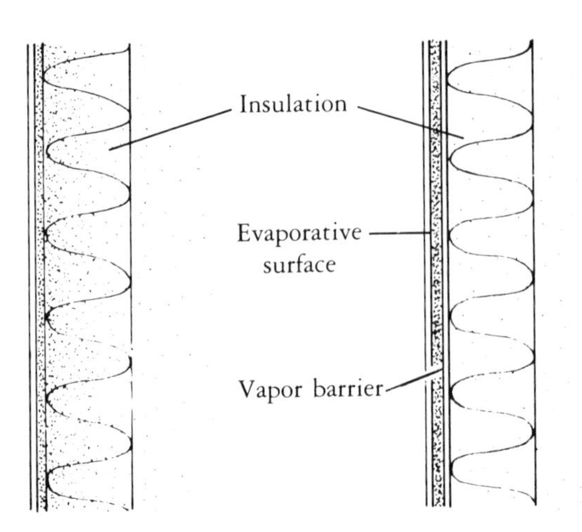

Fig. 31. How vapor barriers reduce heat loss. Diffusion of water vapor along a concentration or vapor pressure gradient (indicated by stippling) takes heat with it—about 2450 joules for every gram of water evaporated at the skin surface. A vapor barrier situated on the warm side of clothing (right) or on the inside of a building wall prevents this diffusion and thus reduces the loss of latent heat.

the vapor diffuses through the insulation only to condense against the cold plastic (liberating latent heat), then heat is lost quickly to the outside by conduction, regardless of our efforts.

WARM BODIES IN COLD ENVIRONMENTS

Physical versus physiological thermoregulation

We are now in a position to estimate the amount of metabolic energy that must be produced just to offset total heat loss during the winter and thereby maintain normal body temperature under cold conditions. To begin with, we will set up a simple word equation that says "Heat In = Heat Out." The "heat in" side of our equation represents energy produced through metabolism of food or fat plus any energy that may be absorbed from external sources (i.e., from sunlight). The "heat out" side of our equation represents the total loss by those processes that we have just discussed. This loss could now be calculated by combining separate equations for conduction, convection, radiation and latent heat exchange. This, in fact, would be the approach taken by the research ecologist who is interested in the relative contribution of each of these modes to total heat loss. However, for our purposes, certain simplifying assumptions can be made that render our task considerably easier and still provide valuable insight into the energetic advantages of the various behaviors we observe during the winter.

The first of these assumptions is that latent heat exchange is relatively small in most homeotherms during the winter, perhaps amounting to 10% or less of total energy loss, and thus can be ignored without serious error. This assumption is based in part on the observation in some small mammals of a 30% or greater reduction in body water turnover during winter.[1] Our second assumption is that the total heat loss from the outer surface of the skin, fur, or feathers by conduction, convection, and radiation is exactly equal to that which is conducted from the core of the animal to the outside surface. Because heat cannot be stored on the surface, no matter how heat loss from the outside is partitioned between conduction, convection, and radiation, the total must equal the amount conducted through the insulating layers from the core of the animal. This latter equality enables us to simplify our heat balance equation by using only the term for conduction to describe total heat loss from the animal (fig. 32). If we now assume that metabolic heat is the only substantial heat input (i.e., that the animal in winter does not absorb significant amounts of heat from its surroundings), then we can write another simple equation,

$$M = kA\frac{(T_b - T_a)}{d}$$

to represent the animal's energy budget, where M is the heat produced by metabolism. This now is the numerical equivalent of our "Heat In = Heat Out" equation.

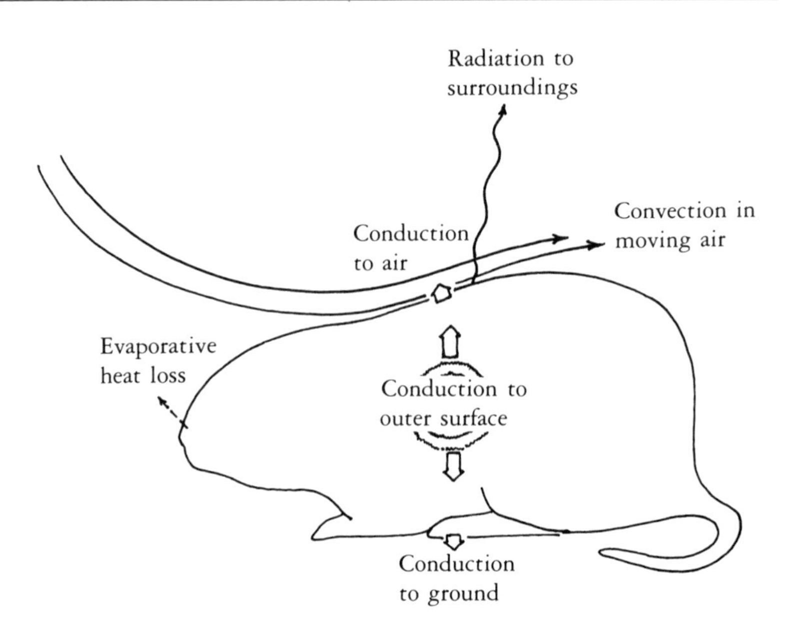

Fig. 32. Heat loss from a small homeotherm. The total amount of heat lost from an animal's skin, fur, or feathers by conduction, convection, and radiation is equal to the amount of heat conducted from the animal's core to the outside of its body. Thus, heat loss in winter can be approximated by using a model for conduction alone, where it is necessary to know only the thickness and thermal conductivity of the animal's insulation, its surface area of exposure, and the temperature difference between the animal's core and its surroundings. Latent heat loss by evaporation is assumed to be negligible at this time of year.

The advantage of this simple model is that we can use it to understand conceptually much of what physical temperature regulation is all about in the winter-active animal. The model tells us that there are several ways the animal can maintain a constant metabolic rate when faced with decreasing ambient temperatures. As air temperature drops further below body temperature (the quantity $t_b - t_a$ increases), metabolism can be held constant by either decreasing thermal conductivity (k), decreasing surface area of exposure (A), or some combination of the two. Also, the thickness of the insulating layer (d) can be increased behaviorally by erection of hairs

THE ENERGETIC ADVANTAGE OF HUDDLING

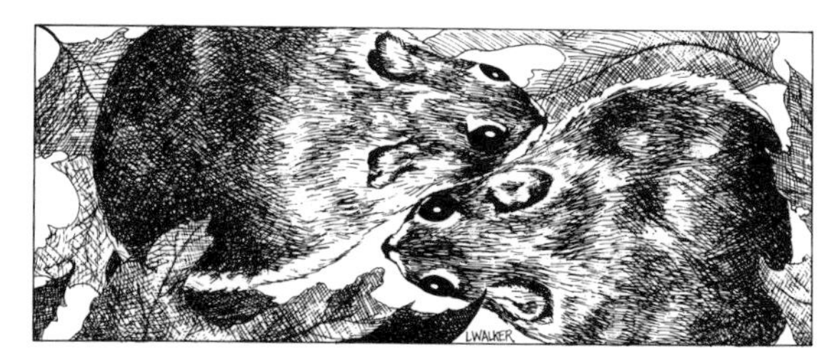

Because heat loss by conduction or convection varies in direct proportion to the amount of surface area exposed by the animal, the energy savings of huddling can be calculated easily. In the case of the small mammals pictured here, assume the following:

- Their effective surface area without huddling is 53.5 cm^2
- The thickness of their insulation (fat, skin, and fur) is .5 cm
- The thermal conductivity of their insulation is .004 W/cm/°K
- Their body core temperature is 37.5° C while the temperature of their surroundings is −20° C
- Their respiratory heat losses are negligible

Heat loss by conduction may now be calculated as follows:

$$Q_c = kA \frac{(T_b - T_a)}{d} = (.004)(53.5)(57.5/.5)$$
$$= 24.61 \text{ joules/second (or 24.61 watts).}$$

Now assume that by huddling each animal reduces its surface area of exposure by one-third. Their heat loss would then be reduced by an equal amount, as follows:

$$Q_c = (.004)(35.68)(57.5/.5) = 16.41 \text{ joules/second.}$$

In this same manner, the energetic advantage of nest building may be calculated by accounting for the effect of the nest on $T_b - T_a$, the body-to-air temperature difference, or by adding the thermal conductivity and thickness of the nest materials to the *k* and *d* terms, respectively.

or fluffing of feathers, as we so often see with roosting birds on a cold winter day. This has the effect of entrapping more air, thereby considerably increasing the insulative value of the fur or feathers. The surface area of exposure can be reduced, of course, by curling and retracting the extremities as much as possible or by huddling with other animals, and this too has a significant effect on the animal's energy balance. A one-third reduction in surface area, for example, by either of these means results in an equal reduction in heat loss (see box on "The Energetic Advantage of Huddling"). Finally, metabolism can be held constant through a reduction in the body-to-air temperature difference ($T_b - T_a$). This may be accomplished by elevating air temperature as much as possible through microclimate modification, such as nest building, or by exploiting the subnivean environment where temperatures are warmer. In some cases reduction of body temperature is also used to accomplish the same result. Some animals, particularly small birds, may undergo nightly torpor or controlled hypothermia, thereby decreasing their body temperature a few degrees and reducing heat flow accordingly.

Many small mammal species that are solitary during the summer become social during winter, constructing communal nests under the snowpack. Table 5 lists those noncolonial species for which there is presently some evidence of social aggregation during winter. The energetic advantages of such communal nesting are by now obvious. A beaver lodge in winter, occupied by two or more animals, remains well above freezing inside the central chamber, even when air temperatures outside are very low (fig. 33). Nests of taiga voles (*Microtus xanthognathus*) occupied by five to ten individuals have been found

Table 5. Noncolonial small mammals that show a tendency to congregate during winter

Clethrionomys gapperi—boreal redback vole
C. glareolus—redback mouse (European), bank vole
C. rutilus—tundra redback vole
Cryptotis parva—least shrew
Microtus californicus—California vole
M. montanus—mountain vole
M. ochrogaster—prairie vole
M. oeconomus—tundra vole
M. pennsylvanicus—meadow vole
M. pinetorum—pine vole
M. xanthognathus—yellow-cheeked vole, taiga vole
Peromyscus leucopus—white-footed mouse
P. maniculatus—deer mouse

Sources: From lists compiled by S. D. West and H. T. Dubun, "Behavioral Strategies of Small Mammals under Winter Conditions: Solitary or Social?" and D. M. Madison, "Group Nesting and Its Ecological and Evolutionary Significance in Overwintering Microtine Rodents," 267–74, both in *Winter Ecology of Small Mammals,* ed. J. F. Merritt, Carnegie Museum of Natural History Spec. Publ. 10 (Pittsburgh, 1984).

to remain between 7° and 12° warmer than ground temperatures and as much as 25° warmer than air temperatures above the snow. Furthermore, the nests are never completely vacated so that foraging animals always return to a warm nest.[2] In addition to this advantage of microclimate modification, the huddling individuals in the nest further reduce heat loss through effective reduction of surface area. It is not surprising, then, that metabolic rates have often been observed to be lower in individuals from communal nests than for animals tested singly (see numerous references cited in the work of D. M. Madison).[3] An added benefit of communal nesting may be a slight reduction in water loss as the higher relative humidity in the nest chamber reduces evaporation,[4] and this of course translates into reduced latent heat loss.

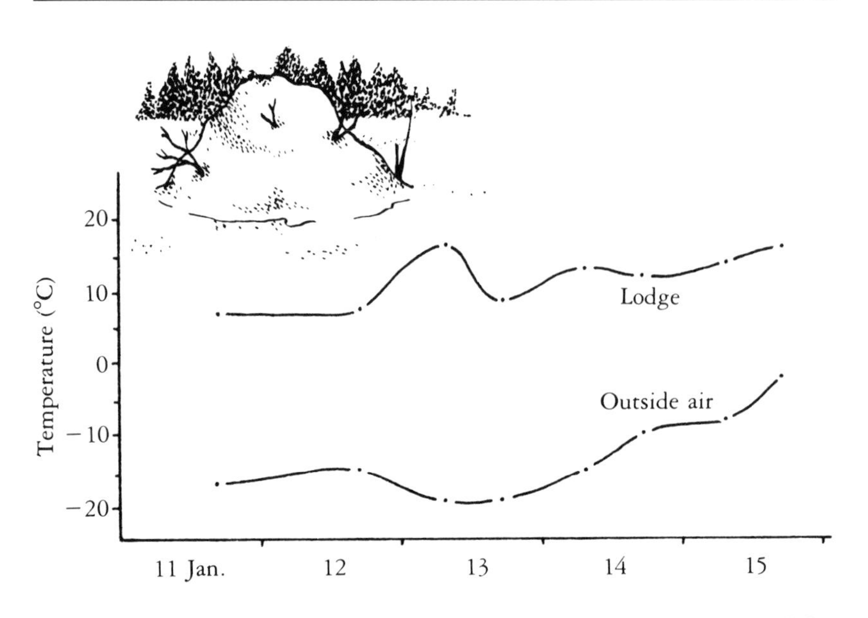

Fig. 33. The ultimate communal nest. The inside of a snow-covered beaver lodge may be as much as 35° warmer than the outside air in mid winter. This lodge was occupied by at least two adult beavers and was monitored by means of a thermistor probe inserted through the lodge wall and into the central chamber.

There are also some disadvantages to communal nesting, such as increased vulnerability to location by predators, a greater possibility of disease or parasite transmission, or more competition for food—but these may be small trade-offs compared to the problems of thermoregulation. With regard to the matter of food competition, communal nesters often show a reduction in body weight that may serve to reduce their total food requirements. Whereas this body size reduction would tend to increase cooling rate at the same time because of an increase in surface area relative to volume or mass, this disadvantage is far outweighed by the energetic advantage of huddling. In essence, communal nesters can well afford to give up body weight if the net result prevents increased competition for food.

The mechanisms that we have seen thus far for maintaining normal body temperature under cold conditions have involved only physical regulation of heat loss. By such behavioral modifications as curling, piloerection (the fluffing of fur or feathers), nest building, and huddling, animals are able to maintain body temperature over an impressive range of ambient temperature fluctuations without alteration of their metabolic rate. There are physical limits, of course, to the extent to which an animal can modify its environment or reduce its own thermal conductivity and surface area of exposure. In the face of still-decreasing ambient temperatures, the animal must at some point increase its metabolic rate in order to balance its heat loss. The temperature at which this becomes necessary is termed the "lower critical temperature" or LCT.

This lower critical temperature varies from one species to another and is seasonally adjusted in many animals, usually through changes in insulation thickness. The red fox, for example, reduces its LCT from 8° C in summer to − 13° C in winter. Likewise, the porcupine adjusts its LCT from 7° C to − 12° C. In both cases this is accomplished primarily through increased thickness of underfur.[5] In the harbor seal, parallel shifts in LCT from 22° C to 13° C in water and from 9° C to −9° C in air[6] suggest that seasonal insulative adjustment is almost entirely the result of an increase in fat thickness. Smaller mammals, on the other hand, are somewhat limited in their ability to add fur or fat. The deer mouse increases the insulative value of its pellage in winter by nearly one-third,[7] but because its fur is still relatively short, the animal's LCT remains quite high. The same is true of the red squirrel, which has the same LCT in winter as in summer.[8]

Some small mammals lower their LCT slightly during winter by a seasonal increase in basal metabolic rate.[9] With a higher level of heat production to begin with, the animal can counter a small increase in heat loss before it becomes necessary to elevate metabolism in order to balance the heat budget. In general, however, the greater the ability of an animal to add fat and fur, the lower its winter LCT. At one extreme, the red squirrel, with a winter LCT of 20° C, is really not very well adapted to the cold and must rely on adequate nest insulation and high food consumption for its survival. At the other extreme, the arctic fox is superbly well equipped to survive extreme low temperatures, being able to remain comfortable at rest to temperatures as low as $-40°$ C without increasing its metabolic rate.[10]

The lower critical temperature, then, represents the point at which the animal has done all it can to maintain a constant metabolic rate through physical regulation of heat loss. Below this temperature, physiological thermoregulation takes over. The animal must now generate additional heat to match that which it is losing to its environment. Because many small mammals have lower critical temperatures well above 0° C, it is clear that metabolic heat production, or "thermogenesis" as it is called, is an important aspect of their winter ecology. Some seasonal adjustment of metabolism may be just as important to their overwintering success as the behavioral adaptations discussed so far.

An increased capacity for heat production without increasing muscular activity ("nonshivering thermogenesis") has been seen during acclimatization to cold in many small mammals. (In earlier discussions of plants and the winter environment the term "acclimation" was used to describe seasonal changes in cold tolerance. In the animal literature, however, this term is generally restricted in use to indicate short-term adjustments, usually under laboratory conditions, while the term "acclimatization" is reserved for long-term or seasonal adaptation.) This seasonal change is accompanied by, and is no doubt related to, an increase in brown fat deposits usually concentrated in the interscapular region of the animal, close to the vital organs. Brown fat is characterized by having a greater number of mitochondria and is thus capable of a higher rate of oxygen consumption and heat production than white fat. It also has a much higher density of nerve cells and blood veins than white fatty tissue. Brown fat is commonly found in abundance in young animals, including human infants, and was once thought to be only a precursor to white fat cells—"embryonal fat" it was called. But brown fat appears in adult animals, too, and is now known to play an important role in nonshivering thermogenesis. In fact, so effective is this tissue in generating heat that temperatures measured just under the skin over deposits of brown fat are sometimes higher than core temperatures.[11] This is most notable in the case of arousing hibernators, where brown fat acts as the heat generator for rewarming the animal.[12] An increase in brown fat shows up in red-backed voles beginning in early fall and continues to accumulate into winter, accompanied by a dramatic increase in the capacity of individuals for nonshivering thermogenesis.[13] These changes are cued by decreasing day length and amplified by low temperature; it is likely that the same mechanism is responsible for the increased capacity for thermogenesis seen during winter in white-footed mice, deer mice, Djungarian hamsters, prairie voles and shorttailed shrews.[14] Apparently the rise in heat production in brown fat upon demand is regulated by the hormone noradrenaline, which is secreted by the adrenal gland directly into the blood stream (fig. 34).

The capacity for producing heat through nonshivering thermogenesis also has its limits. When this mechanism is no longer sufficient to keep up with heat loss, then shivering takes over. For mammals, heat production by involuntary shivering is a last resort, but even this may be considered an adaptive mechanism where the capacity for shivering thermogenesis is subject to seasonal adjustment. In some northern mammals an increase in myoglobin (a protein similar to hemoglobin that binds oxygen) is seen in winter. The increased production of muscle myoglobin observed in red-backed voles during fall acclimatization parallels the seasonal increase in maximum total metabolic capacity in this species and is thought to facilitate oxygen transfer and storage, thus increasing the efficiency of shivering heat production by muscle tissues.[15]

With overwintering birds, thermoregulation takes on a very different pattern than that which we have seen in small mammals. While the same principles of energy exchange apply, the metabolism and behavior of birds is sufficiently different as to require a more physiological approach to regulation of body temperature, even at relatively warm air temperatures. Put another way, oppor-

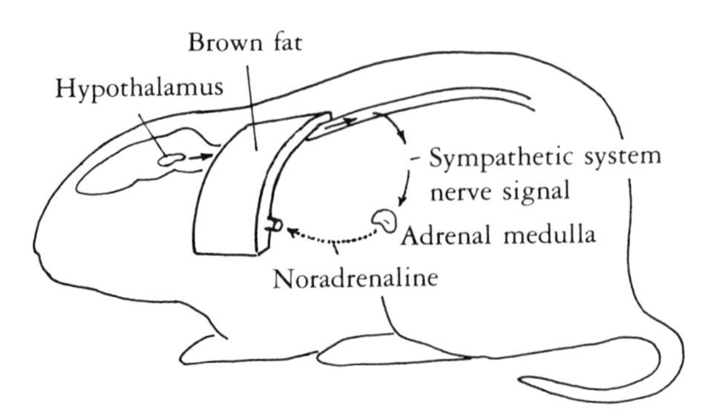

Fig. 34. Heat on demand from brown fat. Brown fat may be likened to a heating pad. As depicted in this schematic diagram, a nerve signal sent from the hypothalamus in the brain (the thermostat) to the adrenal medulla, a gland located next to the kidney, triggers the secretion of noradrenaline directly into the blood stream. It is this hormone that apparently stimulates rapid heat production in the mitochondria-rich brown fat tissue, which is usually concentrated in the interscapular region, close to the vital organs.

tunities for reduction of heat loss by physical means are much more limited among birds. A bird is restricted in its ability to reduce surface area of exposure by changes in posture, and behavioral mechanisms such as huddling are generally not used. Even the ability of the bird to add fat or feathers is limited by the mechanics of flight. Many year-round residents show both a seasonal and daytime increase in fat reserves during winter (see numerous references to this cited in the work of Nolan and Ketterson),[16] but generally birds add little more than enough fat to get through one long, cold night.[17] Nonmigratory dark-eyed juncos in Michigan, for example, are about 14% heavier during winter than their counterparts in Alabama as a result of increased fat reserves, but calculations based on overnight weight loss suggest that this extra fat gives the birds only a 16 to 24 hour advantage in fasting over their southern relatives.[18] Though this certainly may make the difference between success or failure in

Fig. 35. Fat and feathers aren't enough. Though fluffing feathers reduces heat loss, the insulation of small birds is not enough to keep them warm when air temperature drops. Lacking brown fat as well, the winter resident must shiver almost continuously to maintain normal body temperature. (Photo by Peter Marchand.)

the overwintering of northern birds, it is clear that the advantage of added fat is a metabolic one rather than a physical (insulative) one.

The addition of feathers in winter may be of less physical advantage than imagined, too. Birds in the subfamily *Cardueline,* represented by the goldfinches and redpolls among others, may increase total weight of feathers in winter by as much as 50% over their summer weight,[19] no doubt increasing their insulation some. Fluffing of feathers while roosting (fig. 35) might then reduce thermal conductivity of their plumage by another 30 to 50% as compared to daytime values.[20] But even so, studies of several species of northern birds indicate that their insulation is barely sufficient to allow maintenance of normal basal metabolic rates at air temperatures 10° below their body temperature.[21]

Birds, unlike the many small mammals that we have considered, also have limited opportunity to modify the temperature of their surroundings. Some, like the willow ptarmigan and several species of northern grouse, make use of the insulative value of snow by diving into the snowpack and tunneling a short distance to form a sheltered roost. Here they may spend anywhere from a few hours to three days at a time, benefitting from a much-reduced heat loss to their winter environment. If the temperature in the snow burrow remains 5° or so warmer than the surrounding snow and perhaps 25° warmer than the air above the snowpack, the roosting bird may profit by as much as a 45% reduction in energy loss.[22] The best most birds can do, though, is to seek the shelter of dense vegetation or the hollows in snow under shrubs, as redpolls often do, thereby reducing net radiant heat loss to their surroundings[23] (because some energy is gained by back-radiation from the cover around them, even though temperatures may be very low).

Instead of maintaining a constant metabolic rate over a relatively wide range of temperatures through physical thermoregulation, as was the case with small mammals, we see in birds a more gradual shift from physical to physiological adjustment at temperatures below 30° C.[24] This means that a curve of oxygen consumption in relation to ambient temperature is parabolic in shape, with no distinct break in the curve where physiological regulation takes over—that is, no well-defined LCT as we saw with small mammals (though there are LCT values occasionally reported for birds in the older literature). Another important difference in this metabolic curve in the case of birds is that the increase in oxygen consumption over its basal metabolic rate (the rate of metabolism while fasting and at rest) at lower temperatures does not come through nonshivering thermogenesis.

Birds lack brown fat and therefore cannot produce heat through nonshivering thermogenesis. Instead, they rely almost entirely on shivering to maintain body temperature. In fact, the available data indicate that even the larger birds like the crow and raven must shiver continuously during the winter when they are not generating heat through the muscle activity of flight.[25] An increased capacity for shivering thermogenesis during winter is seen in many birds and is apparently related to a seasonal increase in the lipid triglyceride (hence the fat increase that was noted earlier), which is also accompanied by a reduced rate of carbohydrate depletion.[26] Thus, it is not an increase in oxidative capacity of muscle tissue that is responsible for the increased thermogenesis in winter (the demands of flight already dictate a high oxidative capacity). Rather, it is an increase in the availability of fuel to keep the shivering muscle contractions going that enables the bird to produce heat longer at lower temperatures. Winter acclimatized goldfinches, when experimentally subjected to temperatures as low as −70° C, can produce heat at a rate of 5.5 times their basal rate by shivering and can maintain normal body temperature for six to eight hours at temperatures below −60° C.[27]

Chickadees show yet another degree of thermoregulation that involves lowering body temperature during inactive periods. This process, very similar to true torpor, reduces the rate of heat loss from the bird by minimizing the temperature gradient between body and air, and also reduces the amount of energy needed to maintain a stable body temperature. By controlling shivering through shorter and less frequent shivering outbreaks, body temperature gradually drops until a particular depth of hypothermia (varying seasonally) is reached. Shivering is then resumed with regular bursts, maintaining a closely regulated hypothermia. In chickadees this response is not dictated or necessitated by declining fat reserves, but is instead induced by decreasing temperatures and is utilized as a primary means of energy conservation. On a cold winter night the bird may allow its body temperature to drop 10° to 12°, resulting in a savings of 20% of the energy normally required to maintain body temperature.[28]

If we assume that animals partition their use of metabolic energy in some prioritized way between the many life functions and activities that require energy (e.g., foraging, reproductive activity, growth, maintenance of body temperature), then thermoregulation must take the highest priority, especially during the winter. In the words of one expert, "If a [homeotherm] is not thermoregulating, then it is either hypo- or hyperthermic or is becoming so, and ultimately cannot perform its normal functions; therefore channeling energy to other functions would not be possible in any case."[29] Effective thermoregulation, then, is a prerequisite to the success of winter-active homeotherms. We have seen that maintaining a constant body temperature in the face of decreasing ambient temperatures is a multifaceted problem involving two levels of control, physical and physiological, with the latter incorporating both shivering and nonshivering thermogenesis. These controls are summarized schematically in figure 36 and tabulated for a number of species in the box "Balancing the Winter Heat Budget." In many cases these thermoregulatory mechanisms are so effectively used that the animal is able to remain comfortable under most winter conditions (to temperatures as low as −70° C in the case of some small animals)[30] at an energy expenditure often no greater than that required during summer. Subnivean mammals may even have energy sufficient to support nonessential social interactions and, upon occasion, winter reproductive activity. However, animals that remain above the snowcover, often fully exposed to the icy winds of winter, have still other problems to contend with.

The problem with appendages

If a homeotherm didn't need its legs for mobility, it would be far better off without them in winter. The same may be true of all appendages that present a disproportionately large surface area to much colder surroundings. Imagine the problem of a beaver leaving the relative warmth of its dry, snow-covered lodge and slipping into the icy water for a trip to its food cache. As if the sudden loss of insulation weren't enough of a problem, the beaver's broad, flat tail

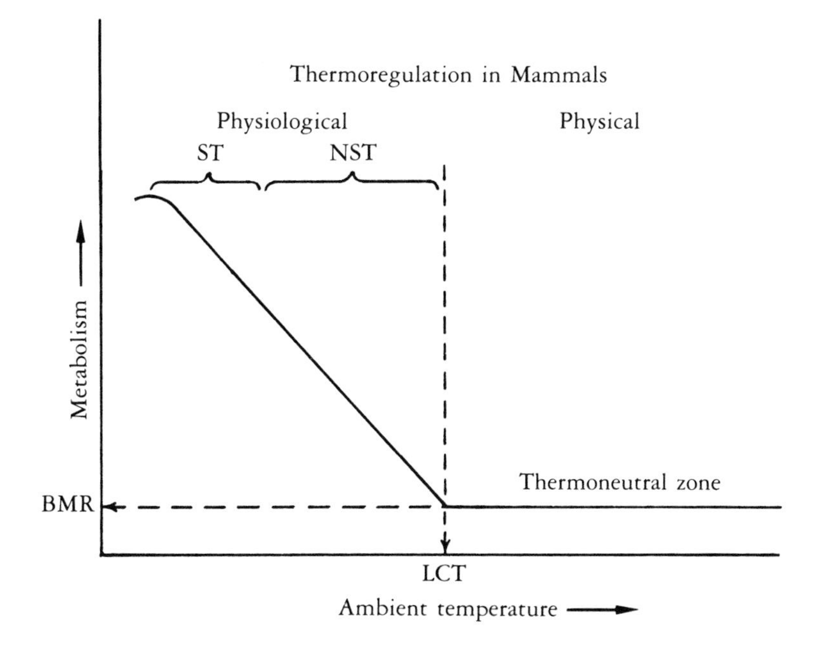

Fig. 36. Physical versus physiological temperature regulation. Within the thermoneutral zone, the animal's resting or basal metabolic rate (BMR) is relatively constant, maintained through physical adjustment of surface area and conductivity, such as by changing posture or by huddling. As air temperature decreases and the limit of physical thermoregulation is surpassed, a point designated as the lower critical temperature (LCT), metabolic rate and consequent heat production increases. In most mammals this stepped-up heat production is accomplished first through nonshivering thermogenesis (NST), which involves the metabolism of brown fat, and then, as a last resort, through shivering or increased muscle activity (ST). In birds the LCT is without clear definition. Instead, there is a more gradual shift from physical to physiological thermoregulation, with increased heat production below about 30° C accomplished entirely by shivering.

BALANCING THE WINTER HEAT BUDGET

A number of thermoregulatory strategies are employed by winter-active homeotherms to achieve the same result—maintenance of body temperature when the mercury drops. The table below is a compilation from the literature on the given species.

	Common shrew (Sorex araneus)	*Redback vole* (Clethrionomys gapperi)	*Prairie vole* (Microtus ochrogaster)	*White-footed mouse* (Peromyscus leucopus)	*Black-capped chickadee* (Parus atricapillus)	*Common raven* (Corvus corax)	*Beaver* (Castor canadensis)	*Whitetail deer* (Odocoileus virginianus)
Physical								
Increased insulation	X			X	X		X	X
Subnivean	X	X	X				X	
Communal nesting		X	X	X			X	
Physiological								
Torpor			No	X	X		X	
Adjusted BMR	X ↑	X ↑	X ↑	X ↑		No		X ↓
Increased NST capacity	X	X	X	X	No	No		?
Increased ST capacity		X			X	X		X
Weight adjustment	X ↓	X ↓	X ↓	No				X ↓
Countercurrent heat exchange							X	

X—known winter adaptation
Blank—not reported in the literature for a given species
No—known not to occur
?—likely, but evidence is inconclusive

in the water seems a perfectly designed radiator, sure to drain heat from the body faster than the animal could possibly generate it. And imagine the problem of a duck swimming in water only a degree or two above freezing, paddling with large webbed feet that seem created as much to dissipate heat from the body as to propel the bird; or the dilemma of a gull standing on an ice floe, circulating blood through its feet to keep those tissues from freezing and then returning the blood to its core for rewarming after giving up its heat to the ice.

The problem with appendages in each of these cases is that they must be supplied with oxygen and kept from freezing by circulating blood through them, but without allowing them to constantly drain heat from the body by conduction through an exposed surface that is disproportionately large relative to the volume or mass of the appendage. The solution to the problem is a physiological shunting of blood through a heat exchanger that intercepts the heat on its way out and maintains the extremity at a considerably lower temperature than the core, thereby substantially reducing heat loss. The heat exchanger in this case consists simply of veins and arteries in close proximity to each other and works as follows:

When air temperatures become low enough, superficial veins in the extremities constrict, increasing flow resistance and thereby shunting more blood through deeper veins lying in close proximity or in contact with the arteries. As cooler venous blood returning from the extremities flows parallel to the arteries carrying warm blood from the core, the temperature gradient between the two favors a transfer of heat from the arterial blood to the venous blood (fig. 37). In effect, a blood molecule traveling toward the core is exposed along the way to successively warmer molecules traveling in the opposite direction in the artery. With the veins and arteries in close contact, heat is easily conducted between them. The returning molecule continues to warm as it moves counter to a constant stream from the core, eventually reaching near-core temperature itself. The arterial blood, on the other hand, is continually giving up heat as it meets the steady stream of colder molecules

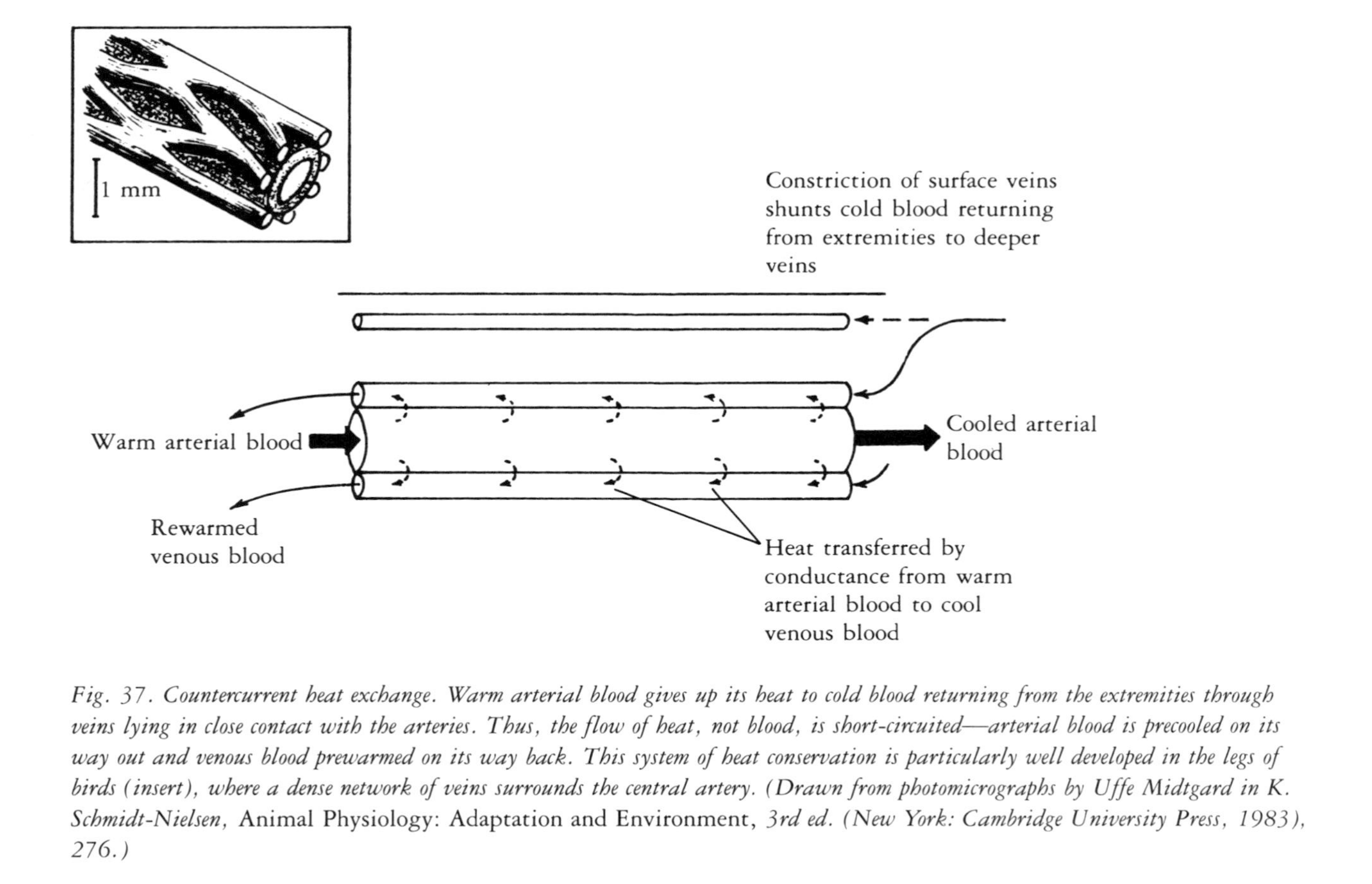

Fig. 37. Countercurrent heat exchange. Warm arterial blood gives up its heat to cold blood returning from the extremities through veins lying in close contact with the arteries. Thus, the flow of heat, not blood, is short-circuited—arterial blood is precooled on its way out and venous blood prewarmed on its way back. This system of heat conservation is particularly well developed in the legs of birds (insert), where a dense network of veins surrounds the central artery. (Drawn from photomicrographs by Uffe Midtgard in K. Schmidt-Nielsen, Animal Physiology: Adaptation and Environment, *3rd ed. (New York: Cambridge University Press, 1983), 276.)*

returning from the extremity. Thus, while the flow of arterial blood reaches the extremities, the flow of heat is short-circuited. Arterial blood is precooled on its way out and venous blood prewarmed on its way in.

This countercurrent heat exchange is especially well developed in some animals, with blood flowing through dense networks of branching and anastomosing arteries and veins in a system known as the rete mirabile ("miraculous net"). Blood circulating through the tail of the beaver passes through such a network and its effectiveness in exchanging heat is dramatic. In air at low temperature, heat loss from the tail may be reduced to less than 2% of the basal metabolic heat produced by the animal; whereas at higher temperatures, over 25% of the heat produced at resting metabolism may be dissipated through the tail.[31]

In whale flippers, each artery is surrounded by veins that precool the flipper and minimize heat loss to the cold water. The same arrangement is found in birds where heat exchange in the legs, particularly of aquatic or wading birds, is critical. An illustration of the artery and anastomosing veins surrounding it in the leg of a European rook is shown in figure 37. It is interesting to note that when the extremities are in imminent danger of freezing, increased blood supply and increased arterial pressure cause an enlargement in diameter of the artery, which then constricts the surrounding veins and forces more blood to return via superficial veins. Thus, an involuntary shunting of blood around the rete in effect delivers warmer blood to the extremities, since heat is not given up to the returning venous blood. When this occurs, the accompanying increase in heat loss to the environment is matched by an increase in metabolic heat production.[32]

BERGMANN'S AND GLOGER'S RULES: IS IT REALLY BETTER TO BE BIG AND WHITE?

These few considerations governing the loss of heat from warm bodies in a cold climate invite other questions regarding the possible adaptive significance of different body sizes and shapes. In a study of the energetics of woodrats and weasels at low temperatures, Brown and Lasiewski demonstrated that the spherical resting posture of the woodrat was far more efficient in reducing heat loss than was the oblong figure of the resting weasel.[33] Their conclusion from this study was that the weasel sacrifices some energetic efficiency as a result of being long and thin, in favor of increased hunting efficiency through its ability to get into small places. But this begs another question often raised in discussions of winter ecology: Is it better to be large or small in cold climates? The often-quoted Bergmann's Rule—that northern races of a species tend to be larger than southern races of the same species—implies that there may be some energetic advantage gained through a decreased surface-area-to-volume ratio. The argument in support of this idea goes as follows:

Consider a cube, an imaginary little heat generator, 1 cm in each dimension. The surface area of the cube is 6 cm^2 while the volume is, of course, 1 cm^3. Hence, the surface-area-to-volume ratio is 6:1. Now double each dimension of this cube. The total surface area increases to 24 cm^2 while the new volume is 8 cm^3. The surface-area-to-volume ratio is now 3:1. Assuming that heat storage is related somehow to volume, and that heat loss is proportional to surface area, then this doubling of size would seem to favor heat conservation by reducing surface area relative to volume.

Admittedly, animals aren't cubes, but arguments favoring an energetic interpretation of Bergmann's Rule follow the same logic; i.e., that a larger animal has less total surface area per unit volume (equating volume with mass) and thereby benefits from a reduced cooling rate (see box on "The Size Trade-Off"). There are two problems with this interpretation, however. First, data on the body size of animals distributed over a wide latitudinal range show mixed results. In general, Bergmann's Rule seems to hold for some large mammals and nonmigratory birds. However, for small mammals, especially those that inhabit the subnivean environment, there is no evidence supporting an increase in body size northward. In fact, as already noted, individuals of some species show a seasonal reduction

THE SIZE TRADE-OFF

Some suggest that a larger animal benefits in winter from a more favorable surface-area-to-volume ratio. Others argue that more heat is lost from the greater surface area and, thus, it is absolute surface area, not relative area, that is important. Which argument is correct? For simplicity in calculation, consider heat loss from the two cubes illustrated here, where both are initially at the same core temperature, but where the dimensions of one are twice those of the other. Given the assumptions stated, it is readily apparent that the larger object loses more heat to its surroundings by conduction; but by virtue of its greater heat content, the larger cube also has more cooling "resistance" (i.e., cools at a slower rate).

5 cm
5 cm
5 cm

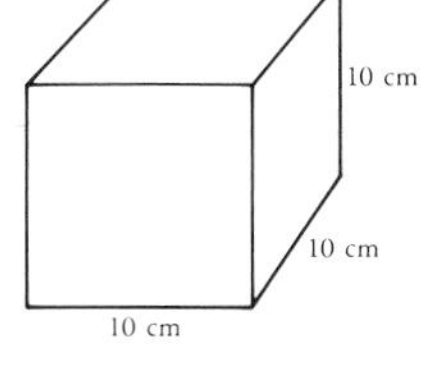

Small cube		*Large cube*
125 cm^3	volume	1000 cm^3
.7 g/cm^3	density	.7 g/cm^3
87.5 g	mass	700 g
2.1 j/g	specific heat	2.1 j/g
150 cm^2	surface area (A)	600 cm^2
37° C	core temperature (T_b)	37° C
− 13° C	air temperature (T_a)	− 13° C
1 cm	thickness of insulation (d)	1 cm
.004 W/cm/°K	thermal conductivity (k)	.004 W/cm/°K
30 W	**heat loss by conduction***	**120 W**
183.7 joules	cooling "resistance"**	1470 joules
16.3° C	**core temperature drop after 100 seconds**	**8.2° C**

*$Q_c = kA\frac{(T_b - T_a)}{d}$

**Heat loss required to lower T_b by 1° (specific heat x mass)

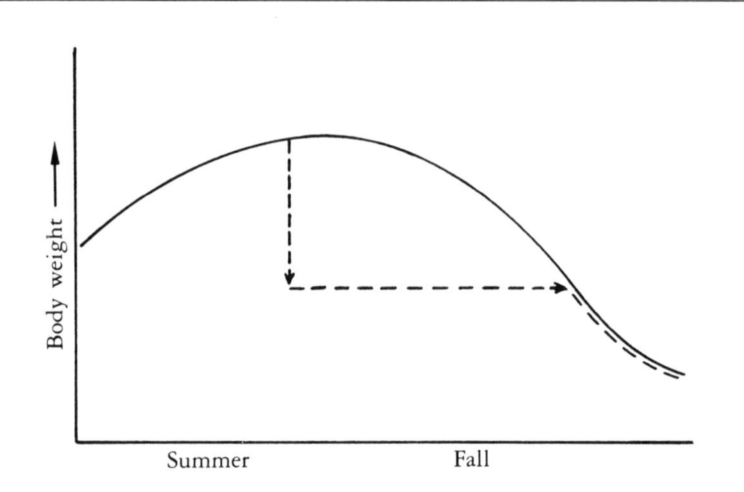

Fig. 38. Weight reduction in preparation for winter. Laboratory hamsters on a restricted diet maintained an artificially low body weight through the summer (dashed line), but still matched the weight loss of their unrestricted cohorts (solid line) in the fall, even when given excess food. (Data from G. Heldmaier.)

in body size as winter approaches. This is apparently genetically controlled, perhaps cued by decreasing day length.[34] G. Heldmaier of Germany placed dwarf hamsters in the laboratory on a restricted diet to artificially reduce and hold their body weight until their cohorts' naturally declining weight reached the same level. The hamsters on the restricted diet were then given surplus food, but instead of gaining weight they keyed into the declining weight curve of their cohorts and commenced losing weight in the same manner (fig. 38).[35] Similarly, free-living voles born in the fall, after only a brief period of growth, key into the same declining weight curve of their earlier spring- and summer-born cohorts, which grow initially to a much larger size.[36]

The second problem with an energetic interpretation of Bergmann's Rule, as argued by Brian McNab, is that animals don't necessarily know anything about their per-gram efficiency.[37] Their only concern is with total food requirements, and the larger animal needs more food, a decided disadvantage during long winters in the North. As an alternate explanation where Bergmann's Rule does seem to hold, McNab suggests that the increase in body size northward is the result of a release from competition. He supports his argument with data showing that the body size of the shorttail weasel, among other animals, increases northward only after it extends beyond the northern range limits of the longtail weasel, its closest competitor (fig. 39). This appears to be the case with other close competitors, too, as for example with the fisher and marten. In western Canada the marten shows an increase in size only north of about 62° N latitude, the range limit of the larger fisher; but in Labrador, where the fisher drops out at 52° N latitude, the marten shows a corresponding size increase immediately.[38]

It is possible, then, that size differentiation is not a primary adaptation to optimize heat balance, but is instead related to prey size selection and competition among similar species for limited food resources. In the case of voles that lose weight at the beginning of the winter, the reduced food requirement appears to outweigh the disadvantage of an increased cooling rate from the smaller body. This is especially true with the communal nester that can afford to lose weight, since huddling offsets the energetic disadvantage of smaller size.

The issue of coloration as it relates to energetics is a bit more complex. It is often suggested that black coloration, because it absorbs more sunlight, also radiates more heat and would, therefore, be disadvantageous to a homeotherm in cold climates. A corollary of this argument, that white radiates less heat, is sometimes suggested as the ecological basis for Gloger's Rule, which tells us that pigmentation in animals tends to be reduced towards the poles, with northern races of animals generally being lighter in coloration than their southern counterparts.

That a black object should radiate heat energy with greater efficiency than a white one would be true only if the black object were heated substantially more as a result of greater absorption of sunlight (assuming the two have equal emissivities). Early discussions

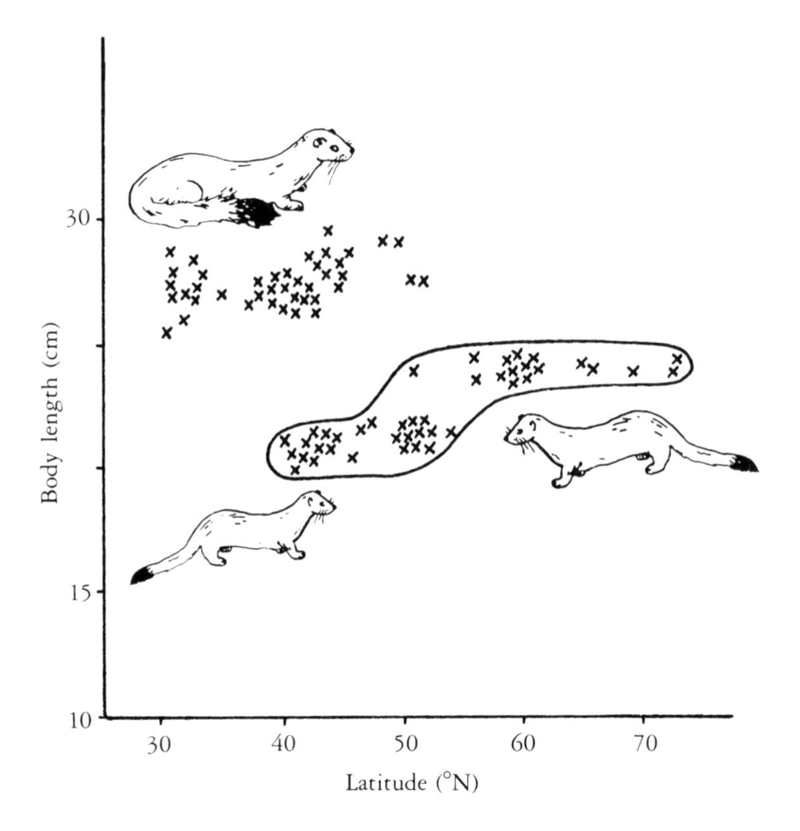

Fig. 39. Body length of male weasels, Mustela frenata *(top) and* M. erminea *(bottom), related to latitude. (Dàta from B. K. McNab, "On the Ecological Significance of Bergmann's Rule,"* Ecology, *52 (1971): 845–54.)*

noted that radiant heat loss from any object is largely a function of its temperature. However, the homeotherm, by definition, maintains a constant body temperature regardless of whether it is white or black. The animal with black coloration may warm more by absorbing solar radiation, but it compensates for this heat load by regulating other modes of heat production or dissipation so as to maintain a constant core temperature. Therefore, it loses no more heat by radiation than does the white animal. This balance between

Fig. 40. The energetic advantage of being black. Zebra finches dyed black metabolize at a considerably lower rate in daylight than do all-white birds because of their greater absorption of sunlight. The use of external sources of heat to balance an animal's energy budget reduces the requirements for thermogenesis. (Data from W. J. Hamilton, III, and F. Heppner, "Radiant Solar Energy and the Function of Black Homeotherm Pigmentation: An Hypothesis," Science, *155 (1967): 196–7.)*

heat absorption and metabolism has been demonstrated under controlled conditions by comparing the metabolic rates of some all-white zebra finches with that of the same finches dyed black (fig. 40). With the "sun" turned on in the experimental chamber, the all-black birds metabolized at a considerably lower rate than the white birds, thus enjoying some energetic advantage from the absorption of light.[39]

This being the case, one wonders then why it isn't of greater advantage, from a purely energetic point of view, to be black in the North rather than white. We think immediately of the importance of cryptic coloration, but what is the advantage to the arctic fox, a true scavenger with little worry about predator avoidance, of being white rather than black? It is interesting to note that in the Pribilof Islands, and in many coastal areas of Alaska and Canada, it is the blue phase of the arctic fox that is most common. If it is important for some reason that the arctic fox blend in with its surroundings, then why do these animals complicate their lives by remaining slate-blue in color throughout the winter? And what is the advantage to the shorttail and least weasels of turning white in winter, when they spend most of their time hunting small mammals under the snow, where visual clues are relatively unimportant to predator or prey? If cryptic coloration were important to the hunting success of predatory animals, then one might expect that the red fox or the weasel's larger relative, the fisher, who hunts above the snow, would turn white in winter, instead. And why does the arctic hare retain its conspicuously white coat all summer on Ellesmere and Baffin Islands and in Greenland?

In an effort to answer some of these questions concerning adaptive coloration in homeotherms, W. J. Hamilton, III, studied the physical and behavioral characteristics of 27 species of all-white and 39 species of all-black birds in western North America.[40] Among his findings he noted that, in general, the insulation of the all-white birds was far superior to that of the black birds, and that this difference was reflected in their roosting characteristics. All of the white birds spent the night in the open, whereas, with only few exceptions, the all-black birds roosted under cover, where they derived some benefit from back-radiation of heat from their surroundings. Hamilton suggested from these data that in birds, at least, white coloration serves to minimize heat exchange with the environment. They reduced loss through increased insulation while at the same time reducing heat gain through increased reflectance, which helps to prevent overheating when the animal is active during the day.

Is the increased insulation of the all-white birds in Hamilton's study merely coincidental, or is there a relationship between color and insulation? Whiteness in most terrestrial animals is due to the absence of the granular pigment melanin. When lacking melanin, the hair of mammals is highly vacuolate (hollow), with the result that light is scattered and reflected rather than being absorbed, and thus it appears white. The same phenomenon is responsible for the reflectance of white feathers. The fact that the white hair or feather barbule is hollow has another important implication. Leaving air spaces in place of pigment granules surely reduces the thermal conductivity of the pelage or plumage. Thus, there may indeed be a physical reason for the better insulation of the white birds and for the superiority of the white arctic fox's insulation over other animals with fur of equal thickness.[41]

Unfortunately, there are no conclusive data in the literature to help resolve our questions here. Even the stimulus for color change in northern animals gives little clue as to the real advantage of white coloration. In some species, temperature appears to control the seasonal change in color, but in others, even within the same genus, photoperiod may be of overriding influence. Experimentally, both short days and low temperatures, by themselves, have been found to induce a change from summer to winter coloration, while exposure during winter to long days for some animals or high temperatures with others can reverse the process. In either case, under natural conditions both temperature and photoperiod would serve to time the molt so as to coincide with annual changes in snowcover. Thus, we can't rule out the possibility that snowcover is still the primary selective pressure that led to the evolution of seasonal color change. Perhaps increased insulation is a fortuitous secondary advantage. We can only agree that there is more to white coloration in winter than meets the eye, and that the possible energetic advantage of having white fur or feathers remains an interesting question in the winter ecology of birds and mammals.

THE COLD-BLOODED GAMBLE: TO FREEZE OR NOT TO FREEZE

For the homeotherms that remain active year-round in cold climates, the principal challenge of winter is balancing the heat budget. For insects and other cold-blooded animals (poikilotherms), winter survival is a different story. While a few of the winged insects are endothermic (having the capacity to elevate body temperature by their own metabolism) and may even be active at near-freezing temperatures,[42] most insects are unable to produce any significant quantity of heat in winter. Even those that do manage to warm above ambient temperatures, either metabolically or by absorbing heat from external sources (e.g., sunlight), have difficulty maintaining elevated temperatures for any length of time because of their small size and rapid heat dissipation. With some exceptions, then, insects, like plants, remain in thermal equilibrium with their surroundings through much of the winter. Many will benefit from the more benevolent environment of the subnivean space; others will seek warmer microclimates, entrenching deeper into the soil. Some, like the snow scorpionfly (*Boreus brumalis*), will even emerge occasionally at the surface of the snowpack (fig. 41*a*), moving about slowly in seeming temptation of fate, while their body temperature warms several degrees above air temperature by absorption of radiation.[43] But come late afternoon, they too slip back into the depths of winter and into thermal equilibrium with the snowpack. Thus, a great number of insects will remain exposed to the cold throughout winter, many with little more than a curled leaf or hollow plant stem for protection (fig. 41*b*). For these, subfreezing temperatures and, on occasion, rapid fluctuations of body temperature across the freezing point of body fluids, are a normal occurrence.

Insofar as insects often experience the same winter conditions as the plants on which they may seek refuge, it seems appropriate to raise some of the same questions we asked regarding plants and the winter environment. Do insects acclimatize to acquire cold hardiness just as plants undergo acclimation? If so, does this process require or involve an altered state of metabolism? Is the induction of cold hardiness triggered by the same environmental cues as for plant hardening? In short, the answer to each of these questions is "yes." To leave it at that, however, would be to shortchange one of the most interesting stories in winter ecology. For what appears at first

a *b*

*Fig. 41. Different approaches to overwintering in insects. The snow scorpionfly (*Boreus brumalis*), whose winter biology is relatively unknown, remains beneath the snowpack through most of the winter, emerging at the surface briefly when air temperatures climb above the freezing mark (*a*). The goldenrod gall fly larva (*Eurosta solidaginis*) overwinters with only the enlarged plant stem protecting it from the elements (*b*). When air temperature drops to the extreme, so, too, does its body temperature. This insect tolerates intercellular ice formation and survives freezing to −50°C and below. (Photo credits: 41a by J. D. Shorthouse; 41b by Jeffrey G. Davis.)*

a passive resistance on the part of the insect to winter's stresses is in reality a sophisticated biochemical strategy for surviving the lowest of winter temperatures.

The problem with freezing

The hazards of ice formation in insects are much the same as with freezing in plant tissues. Growing ice crystals exclude dissolved substances, causing solutes to concentrate in the unfrozen medium at potentially deleterious levels. In addition, ice crystals themselves seem to cause a mechanical disruption of cellular functions, especially of membranes, and are thought to be lethal when they form inside cells. The one known exception in insects is freezing that occurs within cells of the fat body (comparable in function to our liver), though intracellular ice tolerance in other specialized cells may yet be discovered. For now, it appears that the evolution of freezing resistance in insects consists largely of adaptations that render intracellular contents unfreezable, confining ice to intercellular spaces.[44]

Confinement of ice outside the cell in animal tissues may lead to adverse consequences, however. Unlike the plant cell, the interstitial fluid of animals—the hemolymph in the case of insects—has a high solute content. Altering the concentration of these osmotic substances by freezing (termed "freeze concentration") not only elevates solute levels in the extracellular fluid, but may create a condition of "osmotic shock" through a rapid redistribution of solutes and water across the cell membrane.[45] As previously noted, extracellular ice, by virtue of an energy gradient favoring movement of water toward the ice surface, can cause cell dehydration.

Thus, with the insect, extracellular freezing appears a case of double jeopardy. In plants, the solute concentration inside the cell becomes increasingly higher with extracellular freezing, tending to counteract the effect of the ice and at least forestalling the possibility of cellular dehydration. In insects, an elevated solute concentration outside the cell membrane may amplify the energy gradient established by the extracellular ice, thus contributing to the danger of cell dehydration and the likelihood of protein denaturation or membrane disruption through excessive shrinking. There is also the possibility that molecules in freeze-concentrated interstitial fluid will have altered properties that affect metabolic processes; an increase in viscosity, for example, could change diffusion rates.[46] And finally, the growth of ice intercellularly, especially through the recrystallization of small, thermodynamically unstable crystals into larger ones, may damage capillaries or otherwise disrupt cell-to-cell communication, thereby blocking intertissue transport of such materials as oxygen, fuel, or waste products. As stated by one team of experts, "ice in extracellular fluid spaces [of ectotherms] imposes upon individual cells the compulsory requirement to survive autonomously during freezing exposures."[47]

As with plants, it is ice and not low temperatures per se that poses the greatest threat to the overwintering success of insects. Some can tolerate ice extracellularly and, in fact, display special adaptations to ensure early ice formation in order to avoid the risks of supercooling, but then go to great lengths to control the growth of extracellular ice crystals. Others can tolerate no ice whatsoever and go to opposite extremes to prevent freezing, often supercooling to extraordinarily low temperatures. It appears that most insects fall into one category or the other—either "freeze tolerant" or "freeze susceptible"—though with some interesting variations in biochemical strategy among them.

Freeze avoidance

Phenomena such as rime ice formation and plant freezing demonstrate that very small volumes of water can resist phase change to solid at temperatures well below its normal freezing point, especially if confined in a small space. In the plant, spontaneous nucleation of ice in intercellular spaces often did not occur until the plant had supercooled to $-8°$ or $-10°$ C. It is not surprising, then, that insects, being essentially small containers of water, can supercool also, and often do so readily. In fact, a large number of insects (as well as spiders, mites, and ticks)[48] take advantage of this natural ability, often amplifying it through biochemical means to avoid freezing altogether. These are the "freeze-susceptible" or "freeze-avoiding" insects; they cannot tolerate ice formation in body tissues and must develop an enhanced ability to avoid freezing by supercooling.

In freeze-avoiding insects the first 20° C or so of supercooling may be "free" in the sense that it is a consequence of a low volume of water alone.[49] To survive lower temperatures, however, the freeze-susceptible insect must actively lower the supercooling point through the direct inhibition of ice nucleation or actually depress the freezing point. The former may be accomplished in several ways.

The seeding of ice in insect tissues by contact with external (environmental) ice may be avoided simply through selection of a dry hibernation site, the key to winter survival for some. The waxy outer integument of the insect, which is impermeable to water, also serves as a mechanical barrier against external ice seeding, as would a water-impermeable cocoon for pupating stages. However, any small particle can serve as a nucleus for the seeding of an ice crystal and many such nucleators exist inside the insect, particularly within the gut. These include digestive microbes, food particles, and foreign mineral material or dust. A common aspect of cold hardening, then, often involves evacuation of the gut in the fall when the insect ceases feeding.[50] This might be accompanied by one or more biochemical changes, such as removal of lipoprotein nucleators from the hemolymph, which is thought to be responsible for an 18° C drop in the supercooling point of the stag beetle (*Ceruchus piceus*).[51] Other potential nucleators may be masked within cell organelles or membranes.[52] Additionally, the insect may undergo substantial changes in water content during the hardening process, thereby reducing significantly the total amount of freezable water. For example, in the goldenrod gall moth larva (*Epiblema scudderiana*), extractable water content during cold hardening drops from just under 60% to approximately 25% of the fresh weight of the larva, paralleling very closely the seasonal decrease in environmental temperature and insect supercooling point (fig. 42).[53] Some of this change may repre-

*Fig. 42. Winter survival strategy of the goldenrod gall moth larva (*Epiblema scudderiana*). The supercooling point of the insect (b) closely parallels, but is safely below, the seasonal progression of air temperatures (a). This is made possible in part by a decrease in body-water content (c) and a dramatic increase in glycerol production (d) as winter approaches. (Reproduced from J. Rickards, J. J. Kelleher, and K. B. Storey, "Strategies of Freeze Avoidance in Larvae of the Goldenrod Gall Moth,* Epiblema scudderiana: *Winter Profiles of a Natural Population," Journal of Insect Physiology* 33 (1987): 443–50. Originally published as figure la–d in K. B. Storey and J. M. Storey, "Freeze Tolerance and Freeze Avoidance in Ectotherms," C. H. Wang, ed., Comparative and Environmental Physiology 4: Animal Adaptation to Cold *(Berlin: Springer-Verlag, 1989), 57. Used by permission.)*

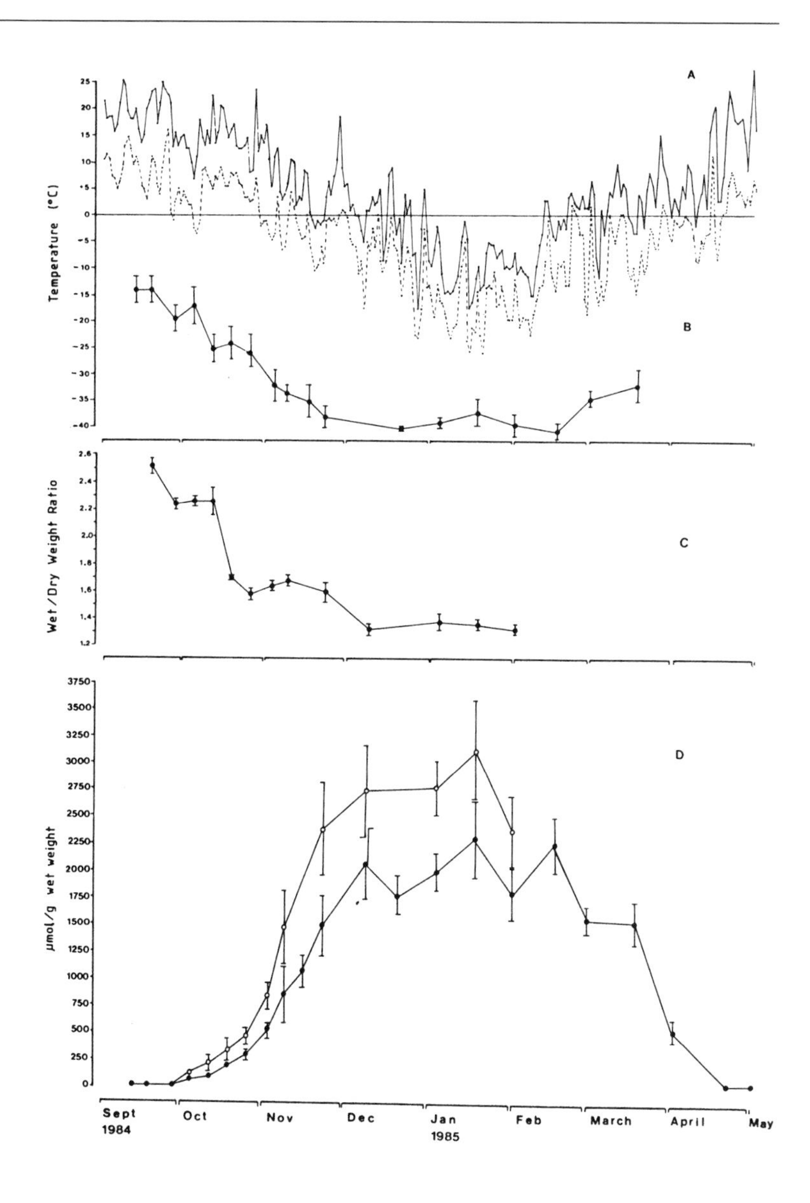
A
B
C
D
Temperature (°C)
Wet/Dry Weight Ratio
μmol/g wet weight
Sept 1984
Oct
Nov
Dec
Jan 1985
Feb
March
April
May

sent hydrolysis of glycogen to glycerol (see following paragraph) or to blood sugars. Or it may reflect, in part, a conversion of bulk water to bound water, in which water molecules become an integral part of some macromolecular structure, and, hence, not necessarily lost from the organism.[54]

In addition to these behavioral, anatomical, and biochemical strategies, some freeze-avoiding insects produce special antifreeze proteins (termed "thermal hysteresis proteins" because they affect freezing and melting points differently). These appear to lower both the supercooling and freezing points, but more importantly, may function to block ice formation by adsorbing onto the surface of embyronic ice crystals, thereby preventing their growth.[55] In this capacity, thermal hysteresis proteins would serve to stabilize the otherwise very tenuous state of the deeply supercooled insect.

Freeze-avoiding insects may also produce varying amounts of low molecular weight compounds, most often polyhydric alcohols (polyols) that serve as additional cryoprotectants, lowering both the supercooling point and freezing point. Glycerol is the most common polyol synthesized during cold hardening, but sorbitol, mannitol, and ethylene glycol may also be found either alone or in combination with glycerol.[56] In the nucleator-free condition resulting from the removal or masking of internal ice nucleators, these polyols are particularly effective in reducing the insect supercooling point. This is perhaps best illustrated in the freeze-avoiding larva of an arctic willow gall insect *Rhabdophaga* spp., where glycerol levels during hardening reach 20% of its fresh weight, while its supercooling point of −66° C is the lowest yet recorded.[57] While there is a general tendency for freeze-avoiding insects with high glycerol concentrations to have lower supercooling points, the correlation is imperfect. The adult arctic Coleoptera *Coccinella quinquenotata,* for example, produces no glycerol, but supercools below −24° C. However, in this insect total blood sugar increases more than threefold during cold hardening, implicating a role of sugars in freeze protection where polyols are not synthesized.[58]

Freeze tolerance

Any way you look at it, deliberate supercooling is a gamble. At temperatures below 0° C, the only stable state in which water can exist is solid. The supercooled insect, then, lives in a metastable state where any disturbance can cause spontaneous nucleation, flash freezing and instant death; the lower the temperature, the more tenuous its existence.

One way to avoid such risk is to promote early and gradual freezing in the extracellular fluid of the insect, and then control the growth of extracellular ice crystals to minimize the problems of extracellular freezing cited earlier. "Freeze tolerance" is a strategy employed by a number of insects in the orders Coleoptera, Diptera, Hymenoptera, and Lepidoptera (as well as marine intertidal molluscs and barnacles, some terrestrially hibernating frogs, the garter snake, and hatchlings of the painted turtle.[59] This, of course, involves a different priority of regulatory functions, though it is apparently within the means of some insects to practice both strategies.[60] The essential requirements of freeze tolerance are the ability to (1) induce ice formation at temperatures only a few degrees below 0° C, (2) restrict ice to extracellular spaces, (3) limit the total amount of body ice and, hence, concentration of solutes, and (4) protect membranes against the problems of structural damage or protein denaturation that accompany freeze dehydration.[61]

It is the ability to induce ice formation prematurely that sets the freeze-tolerance strategy apart. In order to promote early ice formation, the insect has to overcome the tendency of small volumes of water to supercool by producing special ice nucleating proteins in the extracellular fluid. These nucleators are synthesized in the fall, disappear in the spring, and are distinctly different from the non-specific nucleators that are evacuated or masked by freeze-avoiding insects. The mechanism of action is uncertain at present, but it is hypothesized that these proteins somehow reduce the energetic barriers to ice formation, perhaps by providing adsorption sites that rearrange water molecules in a way that facilitates ice lattice formation.[62] In any event, at high subfreezing temperatures (usually above − 10° C) ice formation around these extracellular nuclei proceeds relatively slowly, taking up to 48 hours for maximum accumulation in some insects, which allows more time for cellular adjustment and minimizes osmotic stress.[63] As freezing progresses, water and solute movements out of the cell gradually reach an equilibrium with the freeze-concentrated extracellular fluid, intracellular freezing point is lowered, and the risks of further supercooling within the organism are minimized.

If allowed to grow unchecked, the presence of extracellular ice can, of course, lead to serious cell dehydration. The lethal limit of ice seems to be about 65% of total body water for all ectotherms. In freeze-tolerant insects, as in freeze-avoiding ones, limitation to the growth of extracellular ice is achieved by the synthesis of thermal hysteresis (antifreeze) proteins. Their principal value here appears to be prevention of recrystallization of the small ice masses formed by the ice-nucleating proteins, probably through the same mechanism by which they prevent ice formation in freeze-avoiding insects (i.e., occupying adsorption sites on embryonic ice crystals so as to block the addition of water molecules). In addition to ice-nucleating proteins, freeze-tolerant insects also produce high concentrations of polyols. These polyols effectively bind water and thereby increase the fraction of unfreezable water in the insect, limiting cellular dehydration. They may serve an important function, too, in stabilizing proteins, especially within membranes. Glucose and other sugars can serve in the same capacity as polyols, too, trehalose being especially important in stabilizing membranes by inhibiting their transformation to the gel phase (loss of fluidity) under freeze dehydration.[64]

We see, then, in these two overwintering strategies, a considerable degree of overlap, even a blending of the two in some cases. The Coleoptera larva *Pytho deplanatus,* which often overwinters under the bark of fallen trees, is tolerant of ice formation, but at the same time produces substantial amounts of glycerol and supercools to −54° C, lower than many freeze-avoiding species.[65] The northern beetle larvae *Dendroides canadensis* and *Cucujus clavipes* can even switch their strategy from one year to the next, from freeze avoidance to tolerance and back again.[66] It is not at all clear, either, what controls the choice, as the stimuli for appropriate biochemical preparations are similar in each case. The conversion of glycogen to polyols appears to be induced in all insects by low temperature alone; 5° C seems a common trigger, and optimum production generally occurs between 5° and −5° C.[67] The synthesis of thermal hysteresis proteins, on the other hand, involves more complex interactions between day length and temperature, as these influence specific hormonal actions. The production of ice-nucleating proteins is probably stimulated by the same factors as for antifreeze proteins, though there is some evidence that a 5° C trigger is instrumental here, too.[68]

It is also difficult to say under what environmental conditions one strategy is favored over another. In the goldenrod gall insects *Eurosta solidaginis* and *Epiblema scudderiana,* we see two species exploiting exactly the same microhabitat yet displaying opposite overwintering strategies. In general, however, freeze avoidance is favored where ambient temperatures are not extreme or where daily temperature fluctuations are not excessive, where seeding of ice in the insect body by the presence of external ice is not a threat, or where the organism is periodically active in winter.[69] While this is intuitively sensible, it is not actually known whether freeze avoidance is quantitatively favored under these circumstances. Clearly, though, there are other circumstances where the more stable frozen state would be safer, such as in extreme continental climates where winter temperatures are very low (the lethal limit for freeze-tolerant insects is generally lower) or where disturbance is likely to initiate freezing in the supercooled state.

Freezing in frogs

For one group of cold-blooded animals, the land-hibernating frogs, the choice is clear. These amphibians overwinter beneath a scant

*Fig. 43. The wood frog (*Rana sylvatica*) in a naturally frozen state. Within an hour after thawing this frog resumed breathing, and six hours later its heartbeat was back to normal. (Photo by Evelyn C. Davidson.)*

cover of leaf litter under the snow, where they may often experience subfreezing temperatures. Here their fate is certain; their water-permeable skin virtually precludes any possibility of avoiding inoculative freezing from ice in their surroundings. Having no way to escape the cold and unable to generate any heat on their own, they must acquire some ability to withstand freezing (fig. 43). Four species of land-hibernating frogs—the spring peeper, chorus frog, grey tree frog, and wood frog—have, so far, been identified as freeze tolerant.

The process of freezing in frogs is the same as for freeze-tolerant insects except that frogs have a much greater mass and do not naturally supercool to the extent that insects do. Instead, freezing is normally initiated at about −2° C to −3° C in frogs, and probably for this reason they do not produce "artificial" ice nucleators, as we saw in freeze-tolerant insects. Frogs differ in one respect, too: the most common cryoprotectant seems to be glucose, rather than the polyhydric alcohols (polyols) so common in insects. To date, only the grey tree frog has been found to produce glycerol,[70] though other cryoprotectants may yet be discovered. The effectiveness of glucose in freeze protection of frogs is apparently not related solely to the presence of additional solutes, for its concentration is not high enough,[71] but is likely related to some additional function such as membrane protection.

Perhaps the most unusual aspect of freeze tolerance in frogs is that it is not at all anticipatory in nature. We saw in insect cold hardiness that low above-freezing temperatures, sometimes in concert with decreasing day length, served as a timely cue for the production of polyols and antifreeze proteins, and that processes like gut evacuation and water loss preceded by a safe margin exposure to subfreezing temperatures. In frogs, however, glucose levels show no increase until body temperature drops below the supercooling point and ice actually begins to form in the animal; suddenly then, glycogen in the liver is rapidly converted to glucose and dumped at an extraordinary rate into the bloodstream, increasing blood glucose 200-fold in eight hours,[72] until the frog becomes severely diabetic! Not only would this seem a risky delay in responding to an inevitable danger, but the problem now becomes one of timely delivery of glucose (or glycerol in the grey tree frog) to body tissues while freezing is progressing. This requires effective cardiovascular function under exceptional conditions. Equal to the challenge, the frog shows a remarkable heart rate change in response to freezing. Within one minute of extracellular ice nucleation, heartbeat has been seen to double.[73]

This unexpected cardiac response is associated with a sudden increase in body temperature from the release of latent heat due to rapid ice nucleation. A 12 g frog having a body water content of 78% (by weight) will, upon freezing, release a substantial amount of heat—enough to raise the temperature of its relatively large mass by 2° C, and enough to stimulate considerably greater cardiac output.[74] By monitoring the electrocardiogram (EKG) of frogs during supercooling, J. R. Layne and his associates found that both body temperature and heart rate peaked quickly after spontaneous ice nucleation; as much as five hours later, though, heart rate was still significantly higher than the initial, prefreezing rate (fig. 44). Even with 40% of its body water frozen, the heart continued to beat at two-thirds of its prefreezing rate. Finally, 20 hours or less after nu-

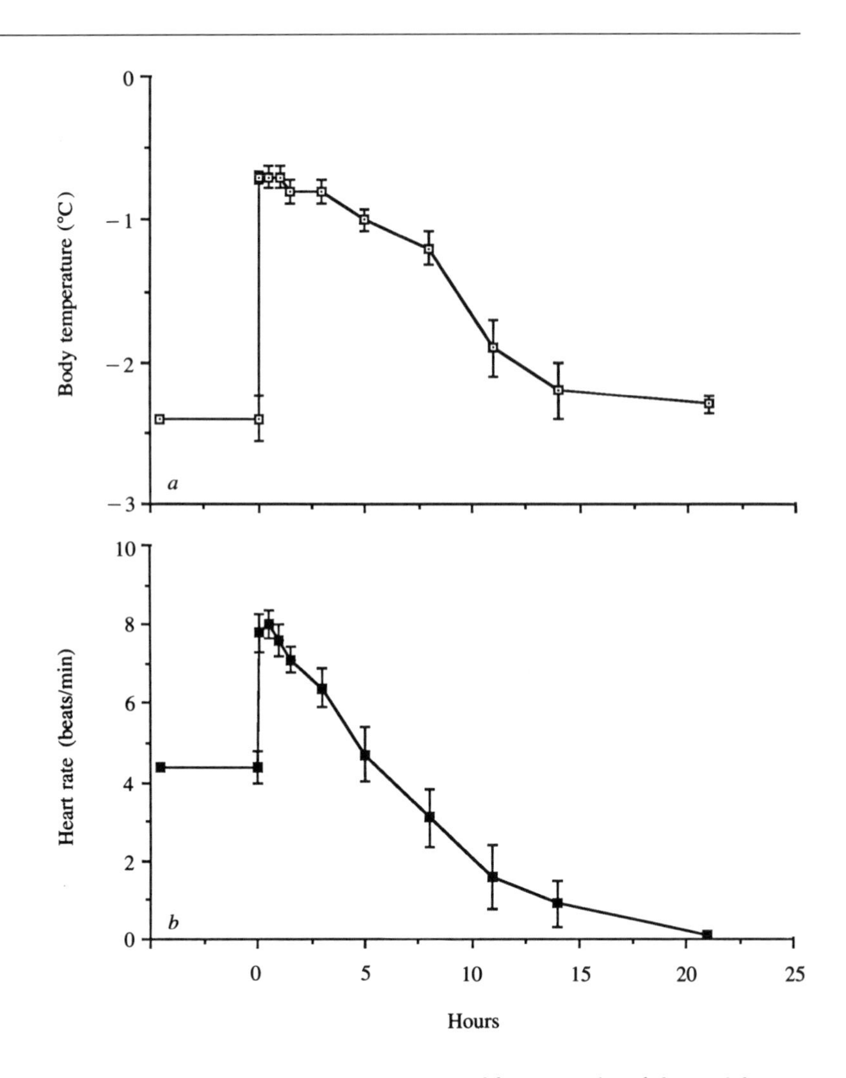

*Fig. 44. Changes in body temperature (*a*) and heart rate (*b*) of the wood frog (*Rana sylvatica*) undergoing nonlethal freezing. The release of latent heat with sudden and rapid ice nucleation stimulates a dramatic increase in heart rate that lasts for several hours and is instrumental in distributing the cryoprotectant glucose throughout the freezing body tissues. (Reproduced from J. R. Layne, Jr., R. E. Lee, Jr., and T. L. Heil, "Freezing-Induced Changes in the Heart Rate of Wood Frogs (*Rana sylvatica*)," *American Journal of Physiology *257 (1989): R1046–49.)*

cleation and with an ice content of 60 to 65%, the heart stops, breathing ceases, and the frog teeters on the very edge of life.[75] It is kept alive now by the almost imperceptible anaerobic metabolism of its energy stores and by the cryoprotectant action of glucose (or glycerol). In its usual hibernaculum under the leaf litter, the margin between ambient (subnivean) temperature and its lower lethal limit is a slim one, for no frogs have yet been found to survive below −7° C.[76] Once again, the importance of a good snowcover seems paramount.

For the frog that overwinters successfully on land, life picks up quickly once temperatures rise again. Within an hour after thawing, the heart resumes beating, and six hours later, at a temperature of only 5° C, heart rate may be back to normal.[77] Thus, the terrestrially hibernating frog can slip in and out of the frozen state quickly. Insofar as these frogs often emerge from hibernation early in the spring to breed, when daily temperatures are still fluctuating across the freezing point, this adaptation would seem to serve them well.

Cold shock

There is yet one more aspect of cold tolerance in insects to consider here. Occasionally, cold-hardy insects suffer injury or death when exposed to subfreezing temperatures that are well above their supercooling points. This is a phenomenon known as "cold shock," and it usually occurs during rapid chilling. Unlike the quick-freezing injury we saw in winter-acclimated plants, cold shock in insects occurs without ice formation. Cold shock has been observed in a wide variety of organisms, and its causes are undoubtedly complex.[78] The actual mechanism of injury is thought to be related to membrane failure, specifically to loss of membrane integrity as the lipids undergo a phase transition from the liquid crystalline state to the gel phase,[79] by now a familiar story.

Cold shock might be a major limitation to the success of insects in microenvironments where temperature fluctuates often across the freezing point, were it not for a remarkable ability of some to cold

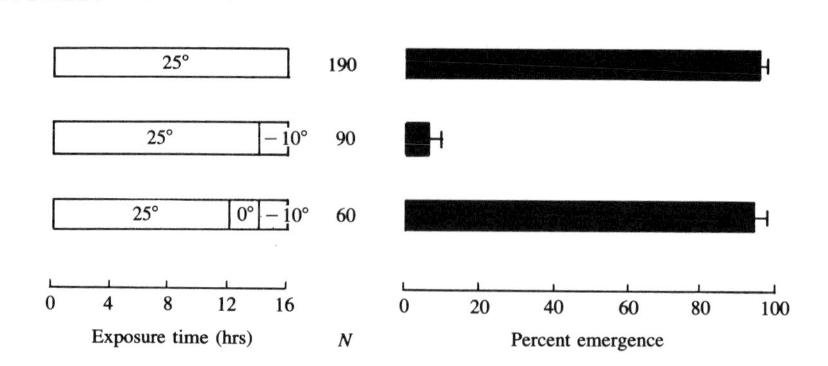

Fig. 45. Cold shock and the effects of prechilling in the flesh fly Sarcophaga crassipalis. *Some nonhardened insects have a remarkable ability to acclimate quickly, producing enough glycerol within minutes of exposure at 0°C to enable them to survive subfreezing temperatures. (Data from R. E. Lee, Jr., C-P. Chen, and D. L. Denlinger, "A Rapid Cold-hardening Process in Insects,"* Science *238 (1987): 1415–17.)*

harden with the briefest exposure to low temperature. Richard Lee, Jr., and his associates have only recently discovered a rapid cold-hardening process in insects, distinctly different from the winter-hardening process. When nonhardened adults of the flesh fly *Sarcophaga crassipalis* are exposed to a temperature of −10° C for two hours, fewer than 20% survive, even though they are capable of supercooling to −23° C. However, if they are chilled first at 0° c for just 10 minutes, half the population survives the −10° C exposure; and if chilled at 0° C for two hours, almost all survive −10° C for two hours (fig. 45). Similar responses were found in other insects and apparently occur at any time of year and in any life stage, even when the insects are feeding and reproducing.[80]

This rapid acclimation is likely associated with an ability to convert glycogen to glycerol very quickly. Indeed, adult flesh flies, after only two hours exposure at 0° C, had increased glycerol levels nearly threefold.[81] In independent studies, isolated fat bodies of silkmoth pupae exposed for just 10 minutes to 0° C showed a doubling of phosphorylase, the enzyme that catalyzes the breakdown of glycogen to glycerol.[82] The activation of phosphorylase at near-freezing temperatures, (unlike most chemical responses in the cold) and subsequent glycerol production undoubtedly confers many of the advantages to the insect that we saw in overwinter cold hardening. At the very least, this rapid cold-hardening process would seem to assure survival through unseasonable cold spells in the fall, until the mechanisms of freeze avoidance and freeze tolerance are in place. With flexibility like this, it is no wonder that these ectotherms can inhabit some of the coldest places on earth.

5 LIFE UNDER ICE

Our consideration of winter ecology began with the arrival of a continuous snowcover on the land, but in the aquatic world it is a different beginning. From the moment late summer nights start to cool, processes are set into motion that soon change the nature of the aquatic ecosystem. As the autumn days crispen and the morning dew turns to frost, one event gains momentum rapidly, an event that will have a profound influence on all life under the coming ice cover. It is the process of overturn—the mixing of surface and bottom waters that redistributes resources in the aquatic environment.

TEMPERATURE/DENSITY RELATIONSHIPS

To understand how this overturn works and why it is so important to organisms under the ice, it is first necessary to understand the unusual relationship between the temperature of water and its density. As warm water cools, it, like most liquids, becomes more dense. Unlike other substances, though, water reaches its maximum density at 4° C. Thereafter, as it continues to cool, water actually becomes lighter, undergoing its most pronounced change when it finally turns to ice. This temperature/density relationship (fig. 46) is a fortuitous quirk, for if water behaved like most other liquids and continued to increase in density as the temperature dropped, then ice would form first on the bottom of the lake, eventually freezing right to the top!.

It is in large part this unusual temperature/density relationship that drives the process of overturn in the lake environment. During the summer, a temperature stratification usually exists in deeper lakes as surface waters are warmed and float above the colder, denser bottom waters. Wind currents may stir this surface layer some, but mixing to any depth is resisted by the density differences that accompany the temperature stratification. Thus, throughout the summer the oxygen-rich waters remain at the surface. However, when air temperature begins to drop in the fall of the year, the stratification that prevented mixing starts to break down. As the surface of the lake cools, the slightly denser water gradually sinks, mixing

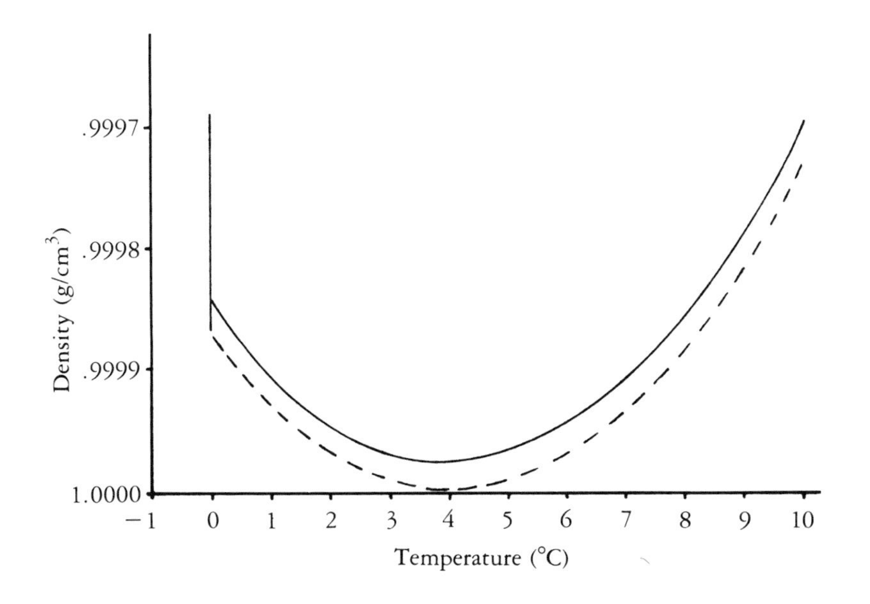

Fig. 46. Temperature/density relationship for ordinary water (solid line) and pure water free of air (dashed line). The density of ice at 0° C is .917 g/cm³.

now with water of lower strata. But the cooler water can sink only to the depth at which it encounters water of the same temperature and density. Colder, heavier water below prevents further subsidence. As the season progresses, however, the continually cooling surface water sinks to greater and greater depths, until finally the whole water column reaches the same temperature from top to bottom. Once this occurs, there are no longer any density differences to resist mixing, and even a gentle wind stirring the surface water can now generate vertical currents that reach all the way to the lake bottom (fig. 47). This top-to-bottom turnover of the water in the fall is an event of significant impact, for it results in the complete circulation of nutrients and oxygen throughout the lake—an important precedent for the hard times ahead. The lake is saturated with oxygen again, and the colder water holds more now than in the summer (the concentration of oxygen saturating water at 5° C is 12.37 ppm compared to 8.84 ppm at 20° C); but this time it is

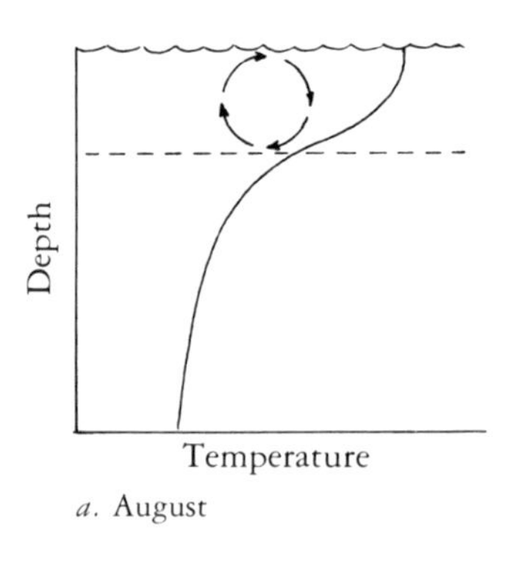

a. August

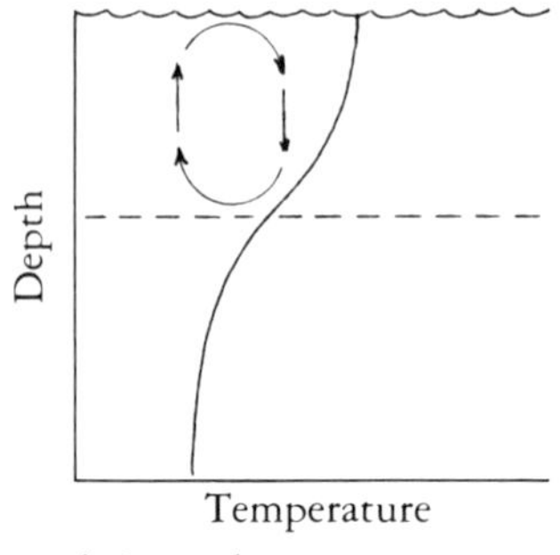

b. September

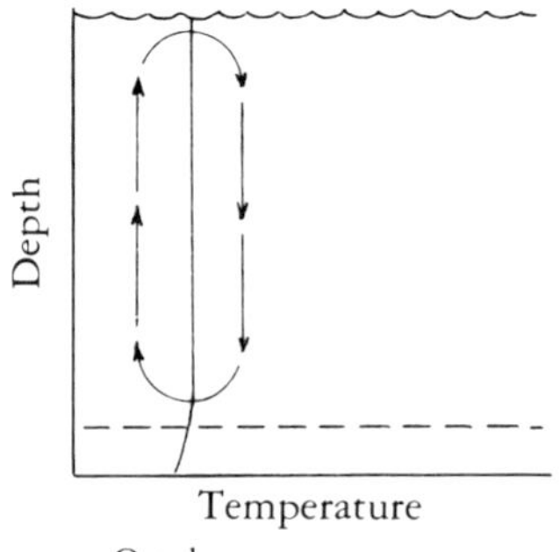

c. October

Fig. 47. Overturn of water as temperature drops. (a) *Warming of surface water results in a steep temperature gradient and stratification of a lake due to differences in the density of water. Mixing of water by wind currents is limited to a relatively shallow surface layer.* (b) *Cooling surface water subsides to a depth of equal temperature, reducing temperature and density stratification in the upper part of the water column and allowing deeper mixing.* (c) *Continued cooling and subsidence has eliminated temperature and density stratification so that mixing currents generated by surface winds now reach to the bottom of the lake. This top-to-bottom circulation is referred to as "overturn" and is important in redistributing oxygen and nutrients before ice cover seals the lake.*

a finite supply, and it may be a long winter under the coming ice cover.

As winter draws nearer, the temperature of the entire water column moves toward the point of maximum density (4° C). Thereafter, water no longer sinks as it cools, but becomes increasingly lighter and stays on top. A reverse stratification develops. The 4° C water at the greatest depth cannot be displaced, so in a lake a few meters deep the bottom water is likely to remain at that temperature throughout the winter. Wind may continue to stir and oxygenate surface waters before freezing, but finally, on a calm night, an ice skim forms and brings an end to the mixing. This is the beginning of winter in the aquatic world.

The thickening ice cover is like a lid on the ecosystem. Light levels diminish with the coming snow, the exchange of gasses between the lake and atmosphere is effectively shut off, and the cycles of aquatic life shift into winter mode. Phytoplankton productivity declines, and the minute zooplankton at the base of the consumer food web drift more freely without encumbrance of the pronounced density gradients that existed in summer waters. Some now even amplify their daily vertical migrations in the water column, responding to exceedingly subtle light cues, while others suspend their usual excursions into the upper waters. With the slowed but continued respiration of many aquatic organisms, plant and animal alike, oxygen levels gradually diminish. Waste products are not recycled as quickly now, and ammonia, carbon dioxide, and hydrogen sulfide levels slowly increase. As winter persists, fish less tolerant of such conditions slowly migrate to inlets or drift to the upper strata of the water. The strain of time in a long northern winter will take its toll on some. If there is any consolation, it is that the aquatic ecosystem is at least buffered against the whims of weather above the ice. Though life for a while is constrained to operating within a temperature range between 4° C, and the freezing point, it is largely without threat of below-zero temperatures that so influenced adaptations in terrestrial ectotherms.

FREEZING AROUND THE EDGES

The exceptions are found at the periphery of the aquatic world. In very shallow waters and running streams, surface ice, especially in times of little snowcover and extreme low temperatures, may extend well into the shoreline or streambank sediments (fig. 48), or even into the bottom mud, as in many arctic ponds. Under these circumstances, benthic (bottom-dwelling) invertebrates will likely experience subfreezing temperatures. In a streambed their exposure may be no more than a fraction of a degree below zero as they become entrapped in anchor ice; but in shallower sediments, temperatures may drop several degrees below freezing, and the benthic fauna must either migrate or develop some freezing tolerance. In an experiment conducted by M. Oswood and his colleagues, natural temperature gradients in shoreline sediments were simulated by placing gravel-filled cylinders in a freezer at $-8°$ C with their bottom portions kept unfrozen (1° C) in a controlled-temperature water bath. Aquatic insects previously introduced at the surface moved slowly downward through the gravel as freezing progressed. Of 154 insects representing 12 different families in six orders, 126 moved to the bottom of the cylinders where, at the conclusion of the experiment, 122 were still alive. Only three of the 28 organisms that remained in the top portion survived the freezing.[1] Evidence from the field for such migrations is largely circumstantial, but points to similar behavior.

A number of aquatic insects do regularly withstand subfreezing temperatures. However, freeze avoidance by deep supercooling, a strategy employed by many terrestrial insects (recall discussion in chapter 4), apparently is not an adaptation of aquatic insects because the frequent presence of ice in their surroundings would likely inoculate freezing in their body tissues. Instead, the available data shows that most aquatic insects supercool only to the same, relatively high temperatures as freeze-tolerant terrestrial insects. Stoneflies, caddisflies, and mayflies are among the more common of the freeze-tolerant insects in winter streams and typically show super-

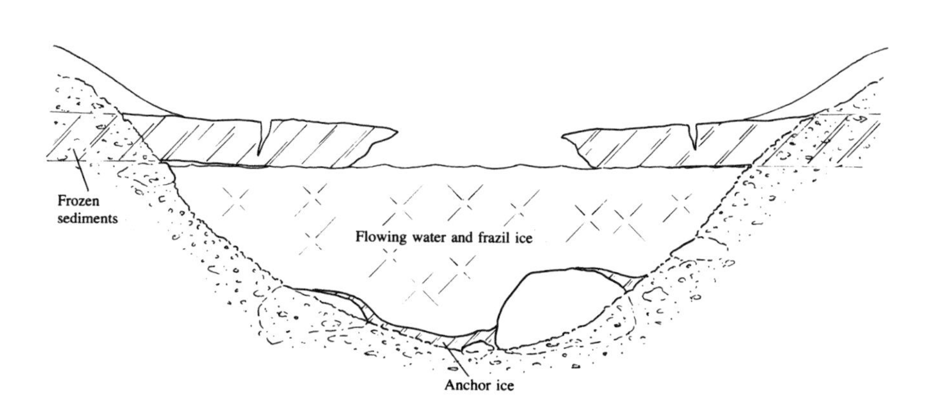

Fig. 48. Profile of a winter stream. In the turbulent flow of mountain streams, water temperature often remains at 0° C throughout the winter. Small disk-shaped platelets of ice known as frazil ice drift in this water, sometimes becoming attached to the stream bottom and growing by accretion to form anchor ice. At the shoreline, surface ice, influenced by subzero air temperatures, may eventually grow outward to span the entire channel, with freezing extending well into the streambank sediments. Surface ice often sags and cracks as stream water level drops throughout winter.

cooling points in the range −3° to −7° C (it is noteworthy, however, that all three orders are usually scarce in frozen sediments, indicating behavioral freeze avoidance). Dance fly larvae (order Diptera) of Alaskan subarctic streams have the lowest reported supercooling point for aquatic insects at −22° C. The latter also exhibit high survival rates after thawing, though length of exposure to subfreezing temperature seems to be of critical importance in the winter survival of these and all other benthic organisms.[2]

In the marine environment, too, we see that some organisms must adapt to subzero temperatures. At higher latitudes, water temperature often drops to −1.9° C, the freezing point of seawater. To a fish whose normal freezing-point depression is only −0.8° C, this poses something of a dilemma, for, in theory, the fish could freeze solid before the water does! For marine fishes of cold regions, then, it becomes necessary in winter either to (1) migrate to warm water or, alternatively, deep cold water where existence in a supercooled state is possible because of the absence of sea ice (remember, the presence of external ice would cause nucleation in the fish), or (2) remain in ice-laden shallow waters and acquire some degree of freezing resistance.

It so happens that a large number of polar and north-temperate fishes fall into the latter category, in which case production of antifreeze molecules depresses the freezing point of their body fluids to just below the freezing point of sea-water. It is a narrow margin of safety, for generally, even in antarctic fishes, artificial exposure to environmental temperatures only slightly below $-2°$ C in the presence of ice results in death. But this modest freezing-point depression is all that is necessary, for unless a body of water freezes to the bottom, the fish will never experience temperatures lower than the freezing point of the water.

In marine fishes, relatively high concentrations of sodium chloride in the body fluids usually account for most of the depression in freezing point. Other salts and metabolites, including potassium, calcium, amino acids, glucose, and urea, generally account for only another tenth of a degree or so depression (fig. 49).[3] These molecules act in a strictly colligative (i.e., concentration-related) manner, interfering, by attraction to water molecules, with the aggregation of water into embryonic ice crystals. Sodium chloride is particularly effective in this regard, as compared to glucose for example, because it dissociates in water into separate sodium and chloride ions, each interacting with the water molecules.

In antarctic and other cold-water fishes, sodium chloride levels in body fluids tend to be somewhat higher, but do not account for the more than twofold increase in freezing point depression. These fishes (table 6), it turns out, all produce antifreeze proteins, identified as linked amino acids (peptides), sometimes with carbohydrate branches (glycopeptides), that are many times more effective in preventing freezing than their concentration alone would suggest. Their primary mode of action seems to be through adsorption onto

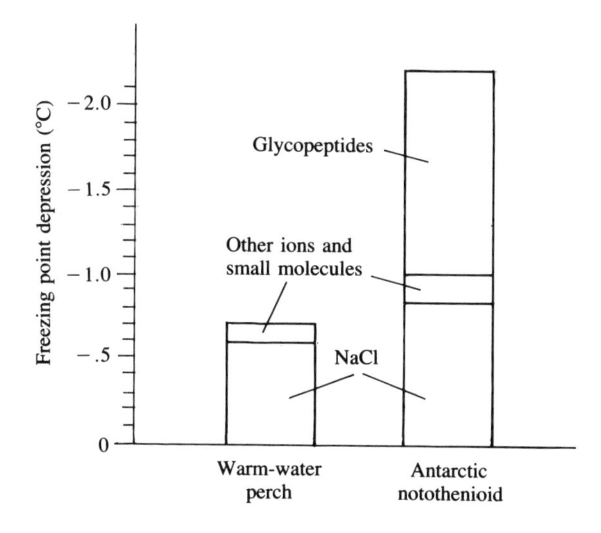

Fig. 49. Contributions of salts and other molecules to freezing point depression in warm-water and antarctic fishes. (Redrawn from J. T. Eastman and A. L. DeVries, "Antarctic Fishes," Scientific American *255 (1986): 109.)*

surfaces of primordial ice crystals and interference with subsequent growth, as we saw with the antifreeze proteins of insects.[4]

In many of the north-temperate fishes, including the Atlantic cod, tomcod, short-horned sculpin, and winter flounder, blood antifreeze levels fluctuate seasonally. The proteins apparently are synthesized in the liver during autumn, in response to declining temperatures and, possibly, shortening day length. In fishes of polar waters where temperature is always close to the freezing point, high levels of antifreeze are maintained in the bloodstream throughout the year, not even diminishing after experimental warm acclimation.[5] This poses an interesting problem from an energetic point of view, because food productivity in cold polar waters is lower than in temperate regions, especially during winter, while the synthesis of antifreeze proteins is energetically expensive. Unexpectedly, however, antarctic fishes show no antifreeze molecules in their urine,

Table 6. Marine fishes producing antifreeze molecules seasonally or year-round

Species	*Geographic region*	*Antifreeze*
Greenland cod, *Gadus ogac*	Labrador	Glycopeptide
Arctic polar cod, *Boreogadus saida*	Bering Sea	Glycopeptide
Atlantic cod, *Gadus morhua*	Newfoundland	Glycopeptide
Saffrin cod, *Eleginus gracilis*	Bering Sea	Glycopeptide
Atlantic tomcod or frostfish, *Microgadus tomcod*	North Atlantic	Glycopeptide
Trematomus borchgrevinki	Antarctic Ocean	Glycopeptide
Pagothenia borchgrevinki	Antarctic Ocean	Glycopeptide
Winter flounder, *Pseudopleuronectes americanus*	Newfoundland	Peptide
Alaskan plaice, *Pleuronectes quadritaberulatus*	Bering Sea	Peptide
Bering Sea sculpin, *Myoxocephalus verrucosus*	Bering Sea	Peptide
Short-horned sculpin, *Myoxocephalus scorpius*	Newfoundland Ellesmere Island	Peptide
Polar eel pout, *Lycodes polaris*	Arctic, Circumpolar	Peptide
Sea raven, *Hemitripterus americanus*	North Atlantic	Peptide
Eel pout, *Rhigophila dearborni*	Antarctic Ocean	Peptide

Source: Compiled from A. L. DeVries, "Antifreeze Peptides and Glycopeptides in Cold-Water Fishes," *Annual Review of Physiology* 45 (1983): 245–60.

suggesting the existence of some mechanism for conserving glycopeptides rather than continuously synthesizing them. Indeed, it turns out that every species examined so far lacks kidney glomeruli, the capillary tubules from which molecules the size of these proteins would be forcefully excreted into the urine. Instead, these fish remove wastes by a secretory process in which cells lining the walls of kidney tubules draw only selected materials from the blood, leaving the antifreeze proteins in circulation. This unusual kidney function may have evolved in these fish specifically as a cold-water adaptation.[6]

DORMANCY VERSUS ACTIVITY: COMPENSATING FOR THE COLD

In freshwater environments the freezing point of the fish body fluids is lower than the freezing point of water, obviating the need for elaborate cryoprotective strategies. However, the challenges of winter in the aquatic ecosystem are by no means limited to freeze avoidance. The aquatic invertebrates and fish that remain active in this environment are all cold-blooded (poikilotherms), which means they must function at body temperatures within the same narrow range as the water. This poses two problems in particular at near 0° C temperatures. First, fluidity of membrane proteins and lipids, especially of the saturated lipids, becomes greatly reduced, impairing membrane functions. Secondly, rates of chemical reactions and the establishment of equilibria between formation and breakdown of reaction products are strongly temperature dependent, tending, in the absence of acclimatory metabolic adjustment, to slow life's processes considerably at low temperatures.

It would not be surprising, then, to find the activities of nonhibernating aquatic organisms greatly limited under winter ice cover. As a general rule, for every 10° C decrease in temperature, the chemically mediated metabolic rate of the poikilotherm could be expected to decrease by a factor of two or slightly more. This means that if a fish were acclimated to a temperature of 25° C and we abruptly put it into water at 5° C, its body functions would slow to only one-fourth the initial rate (one-half for the first 10° and half again for the next 10°).[7] Thus, some aquatic organisms become very inactive at the low temperatures of winter, remaining in a state of dormancy or "semihibernation" in which they are slow breathing and lethargic, but can still respond to physical stimuli. Inactive fish like bluegills and the many species of zooplankton that remain suspended in the water column are aided by the higher density of the cold water that increases their buoyancy and minimizes the amount of energy needed to stay there.

The energetic savings of winter dormancy, at a time when resources may be particularly scarce, is indisputable. In many organisms, the depression of metabolic rate at low temperatures is dramatic. The American eel (*Anguilla rostrata*), for example, shows a twofold reduction in metabolism between temperatures of 10° C and 5° C.[8] This is far greater than would be expected on the basis of chemical kinetics alone and suggests a regulated depression or deactivation of metabolic processes by the organism. The energetic savings in this case is realized through a significantly reduced catabolism of lipid stores, compared to reserve-energy utilization in fasting eels at 10° to 15° C.[9] Very little is known about the mechanisms regulating metabolic depression at low temperatures, but restructuring or dissociation of enzymes from cell organelles and reduced delivery of oxygen to tissues (even though blood oxygen is often higher) are likely possibilities.[10]

Not all aquatic organisms show a depressed metabolic rate as an adaptation to winter conditions. In fact, many poikilotherms show a remarkable ability to adjust their metabolic rate over time as they acclimatize to cold water. As water temperature cools gradually with the approach of winter, the temperature tolerances of these organisms shifts, too, in what is often termed "compensation" of metabolism. Polar fishes commonly show such adaptation to persistently cold water, functioning as efficiently at low temperatures as do temperate-region fishes at considerably higher temperatures. For example, rates of oxygen consumption, protein synthesis, and power output by polar fishes may exceed, sometimes severalfold, that of temperate fishes when measured at near 0° C temperatures.[11]

Similar compensation is seen in organisms of seasonally cold oceans and temperate lakes. The winter-acclimatized poikilotherm in some cases can maintain the same level of activity at a lower temperature as was formerly possible only at a much higher temperature. A salamander, for example, that has acclimatized to a temperature of 15° C might show a metabolic response to rapid temperature changes as indicated by curve *ab* in figure 50. In this case, its metabolic rate is considerably reduced when exposed to

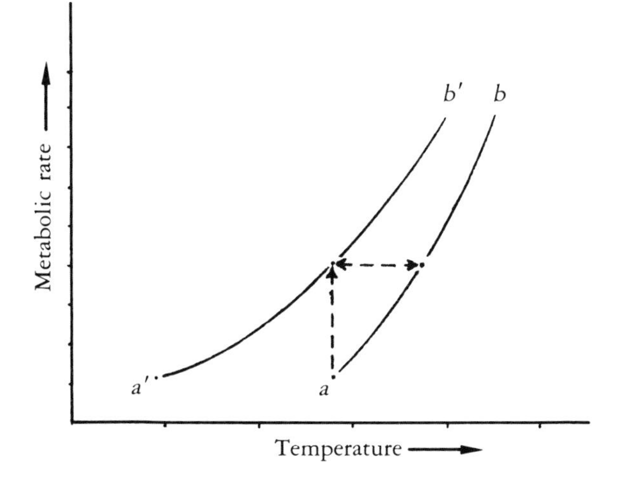

Fig. 50. Upward adjusted metabolic rate in aquatic organisms during winter. The relationship between temperature and metabolic rate in a warm-acclimatized poikilotherm is given by the curve ab. *However, in the process of cold acclimatization this relationship shifts to the left (curve* a′b′*) such that the organism in winter may maintain the same rate of metabolism that was formerly possible only at a higher temperature (horizontal dashed line). Comparing metabolic rates for any given temperature shows an upward adjustment in the cold-acclimatized organism (vertical dashed line).*

colder water. However, if held at the lower temperature for a period of time, its metabolism gradually increases, a response often referred to as an "upward adjusted" metabolic rate. In effect, the whole curve shifts to the left (curve *a′b′* in figure 50), with the salamander soon metabolizing at the same rate as it did previously at 15° C. This adjustment has been observed in red-spotted newts that remained active throughout winter at the bottom of a shallow pond in the North, where a spring kept a small area of water ice-free. A similar response has also been seen to varying degrees in molluscs, crayfish, and sand crabs.[12]

This ability of aquatic poikilotherms to shift to higher metabolic rates at lower temperatures apparently involves changes in enzymatic reaction rates, which can come about in one of two ways. An enzyme that serves as a catalyst for a specific metabolic reaction may exist in several different forms called isozymes. Isozymes are variations of the same basic molecule, but they often exhibit different temperature sensitivities so that one or another controls a reaction at a given temperature. Thus, adjustment to decreasing temperatures may occur through a change in proportions of different isozymes. If, for example, isozyme "*x*" exhibits maximum sensitivity at a temperature of 10° C and isozyme "*y*" shows a maximum response at 5° C, then an increase in the ratio of *y* to *x* would result in greater activity at the lower temperature. In figure 50, idealized for a number of organisms, the convergence of the two curves toward their upper ends suggests such a change in isozyme ratio.

Some researchers feel, however, that the tailoring of isozymes to specific thermal conditions best explains adaptation at the biogeographic level, permitting different populations of the same species to inhabit different thermal environments over a wide latitudinal range. The killifish (*Fundulus heteroclitus*), which is found in coastal waters from the cold North Atlantic to Florida, is a good example. Seasonal acclimatization, these researchers suggest, is better explained by quantitative changes in enzymes, where metabolic adjustment is achieved simply through a change in enzyme concentration—the production of more enzyme to catalyze more reaction. If, in our winter-acclimatized animal, metabolic rates were increased by an equal amount at all temperatures (i.e., curve *a′b′* in figure 50 was parallel to curve *ab*), the implication would be that an increase in total number of enzyme molecules had occurred rather than a change in isozyme ratio.

It is possible, of course, that both qualitative (i.e., shift in isozyme ratio) and quantitative enzyme changes are involved in cold acclimatization. This is suggested by data on pumping rates versus temperature for the North American mussel (*Mytilus californianus*).[13] Populations from latitudes 34°N and 39°N show a parallel metabolic response to declining temperature, but a third population from 48°N latitude exhibits considerably less reduction in pumping rate

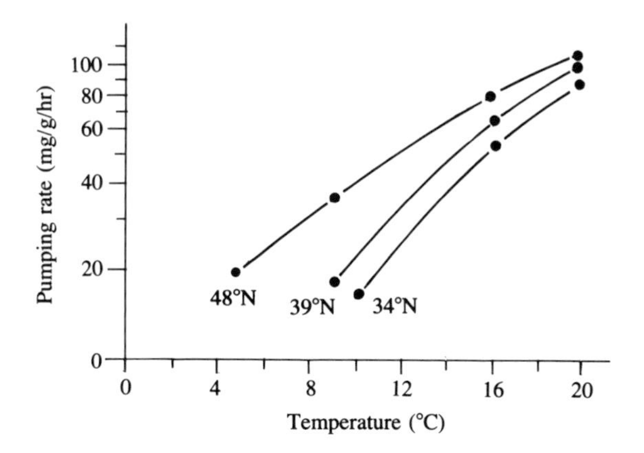

*Fig. 51. Pumping rates of the North American mussel (*Mytilus californianus*) from different latitudes, showing upward adjusted metabolic rates in cold water for the northern populations. The change in slope of the curve for the northernmost population suggests a different mechanism of compensation, likely involving shifting isozyme ratios, perhaps in addition to quantitative changes in enzymes. Compare with figure 50. (Data are from K. P. Rao, "Rate of Water Propulsion in* Mytilus californianus *as a Function of Latitude,"* Biological Bulletin *104 (1953): 171–81.)*

at low temperature, implicating a different strategy in cold acclimatization (figure 51).

Thermal compensation in aquatic poikilotherms may also involve structural modification of tissues and cell organelles, and might include alterations in energy substrate utilization. For example, changes in lipid composition, as we have seen before, may impart greater membrane fluidity and improved functioning at low temperatures. Also, in cold-acclimatized fish, an increase in the amount of muscle red fiber may help maintain swimming performance, and a switch to greater reliance upon lipids for energy demands of muscle tissue at low temperatures may improve power output.[14] These adaptations are summarized in figure 52.

By whatever means it is accomplished, metabolic compensation is clearly of adaptive value in allowing many organisms to function throughout the long winter under ice. But low-temperature accli-

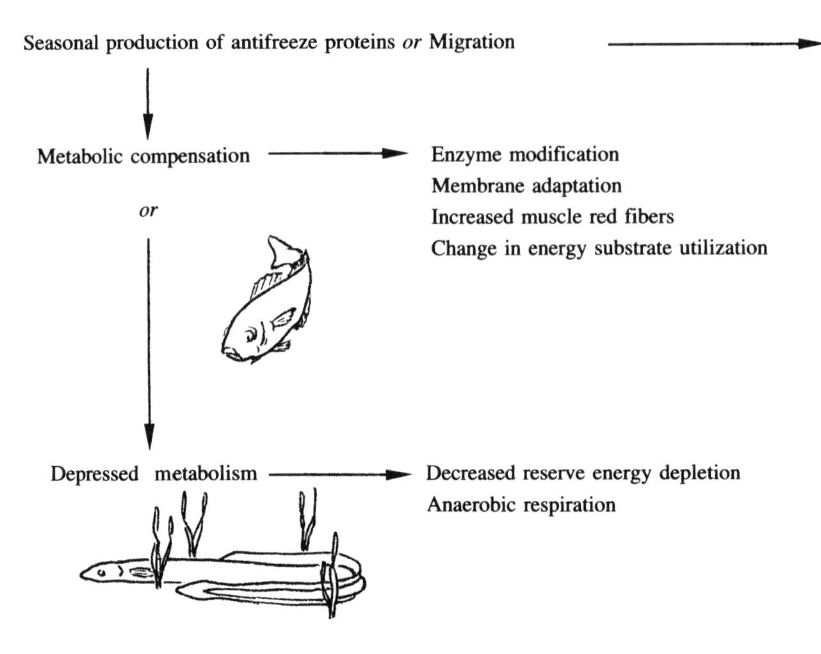

Fig. 52. Coping with cold water. Fishes of both aquatic and marine environments display a number of adaptations to the icy waters of winter. Summarized above, these adaptations generally fall into one of two opposing strategies: a depressed metabolic state or winter dormancy aimed at conserving energy and oxygen, or an elevated metabolic state enabling the fish to remain active at near-freezing temperatures. Only the marine fishes need exercise the first option of antifreeze production or migration.

matization is not the final answer to winter survival, for the organism that remains active faces still other problems. In some lakes it is the progressive depletion of oxygen in the water, rather than low temperature, that creates the greatest stress of winter.

DEALING WITH OXYGEN DEPLETION

All aquatic organisms, including the hibernating reptiles and amphibians in the bottom mud, utilize oxygen throughout the winter. With supply limited now, a gradual deficit accumulates through respiratory consumption of oxygen. The mud-dwelling decomposers are the most numerous and heaviest users of oxygen, so it is usual to see the oxygen depleted first near the bottom as it is removed by these organisms. Oxygen then diffuses downward through the water column toward areas of lower concentration, but only very slowly, giving rise to an oxygen profile or concentration gradient such as that shown in figure 53. It is not uncommon by midwinter to see a near absence of oxygen in the bottom waters. Ice fishermen are well aware of the manner in which many fish species rise to higher strata in the lake as oxygen depletion progresses. As winter wears on, only slow internal currents—perhaps some slight inflow-outflow movements of water—will help redistribute oxygen within the lake.

If the shortage of oxygen becomes severe enough, the less tolerant organisms may die. Fish kills from oxygen starvation are not uncommon in shallow, productive ponds during unusually long winters. When the concentration of dissolved oxygen reaches critical levels, the more resistant organisms may switch to anaerobic respiration in which they derive energy from the breakdown of carbohydrate (usually glycogen) to lactic acid, a process known as glycolysis. This oxygen-saving metabolism is utilized also by hibernating turtles (and perhaps by the amphibians as well) to supply their energy needs throughout the winter.[15] Glycolysis yields only about 8% of the energy that would be available from the complete oxidation of glycogen in aerobic respiration. In some cases, the accumulated lactic acid can still be used as a substrate for later oxidation reactions, thus eventually giving back all of the energy of the original carbohydrate.[16] In some fish that are especially tolerant of oxygen-deficient conditions, the accumulating lactic acid may be converted to ethanol, which then diffuses readily from the gills. This wastes a potential source of energy, but reduces the acid-balance problems caused by excessive lactic acid.[17]

Aquatic plants might be singled out as major contributors to winter oxygen stress but for the fact that some species appear adapted to photosynthesizing at the very low light levels and low water temperatures typical under ice. This, of course, gives back

Fig. 53. Temperatures and dissolved oxygen profiles typical of mid winter in a moderately productive temperate lake 5 m deep. Lakes with lower productivity (fewer oxygen-consuming organisms) will have a more vertical profile, with a higher oxygen concentration in the lower strata.

some of the oxygen used in respiration. Many species of phytoplankton—minute suspended plants including motile algae—remain photosynthetically active throughout winter, shifting in their vertical distribution as light levels change with snowcover conditions on the ice and contributing substantially to the fixation of carbon and release of oxygen in the water.[18] Elodea (*Elodea canadenses*), a common submerged aquatic plant of northern lakes, is known also to increase in biomass under ice,[19] and it appears that much of this wintertime growth is supported by active photosynthesis. Plants that have been gathered through holes in the ice, placed in closed glass containers, and then returned to the pond bottom to "incubate" under normal winter conditions have shown considerable amounts of oxygen evolution as a result of photosynthesis.[20] In fact, in January, under the lowest light levels occurring beneath the ice, these plants produced enough oxygen during daylight hours to pay back 40 to 60% of the oxygen they consumed in respiration over a 24-hour period. At times during January, oxygen production reached 50% of the gross daytime release measured for this species in late summer. This is enough to cause a significant increase, during daylight hours, in the dissolved oxygen concentration of water above a bed of submerged aquatics. Figure 54 shows the daily change in ambient dissolved oxygen attributable solely to vascular plants photosynthesizing under remarkably low light filtered through 47 cm of snow and ice and 1 m of water at a temperature of only 1.2° C. While there is still a net reduction of dissolved oxygen at the end of 24 hours, the daytime pulse is significant. Later in the season, when days are getting longer and light levels under the ice increasing, oxygen evolution increases substantially, with payback exceeding 75% of respiratory consumption. This raises interesting questions about the possibility of daytime migration of other organisms into the shallow littoral zone in response to oxygen production by aquatic plants.

This photosynthetic ability does not exonerate plants from their part in the progressive development of oxygen stress in the aquatic

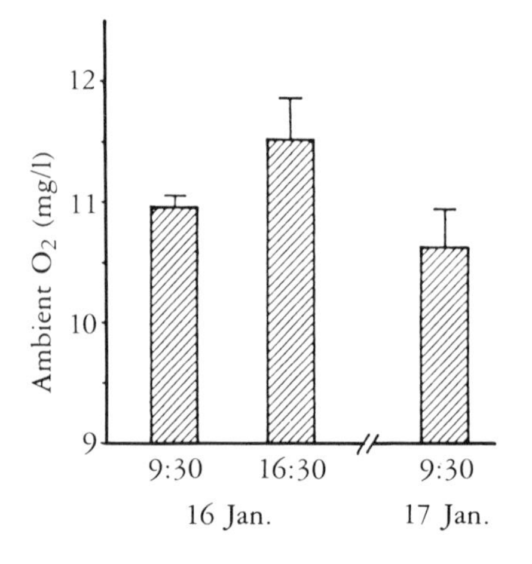

Fig. 54. Daily change in dissolved oxygen concentration in the littoral zone of a shallow lake as a result of photosynthesis by vascular plants alone. The water temperature on January 16 was 1.2° C and maximum irradiance was .61 μE/m/s (.15% of incident above a thickness of 47 cm of snow and ice cover). (Data from P. J. Marchand, "Oxygen Evolution by Elodea canadensis *under Snow and Ice Cover: A Case for Winter Photosynthesis in Subnivean Vascular Plants,"* Aquilo Ser. Botanica, *23 (1985).)*

environment under ice, but without it things might be worse. These plants highlight the remarkable ability of aquatic organisms of all kinds to adapt to the unique stresses of this season. In the end, though, it is a game of endurance, for winters under the ice are very long. A January thaw may bring temporary respite to the land, but seldom does it break the seal on the aquatic ecosystem. Spring melt, too, is slow to release the lake, and even then it is only with the occurrence of spring turnover—a repeat of that fall process that again stirs new life into the deep waters—that winter finally ends in the aquatic world.

6 PLANT-ANIMAL INTERACTIONS: FOOD FOR THOUGHT

Earlier in our discussion of plants and the winter environment, we talked about the mechanical forces of winter—the problems of snow loading, drifting, and ice blast as they affect trees and shrubs exposed above the snowpack. We considered how these forces mechanically shape trees at high elevations, how they affect physiological processes like water relations and carbohydrate balance, and how they influence demographic processes like wave-mortality and the development of ribbon forests in some mountain areas of the western United States. But we did not talk about the ecological impact of concentrated winter browsing by herbivores on woody plants. In areas of the northern coniferous forest where local concentrations of hares or moose are high, or in areas where deer yard for the winter, the browsing of woody plants above the snowpack may constitute a force of major significance in the evolution of plant survival strategies (fig. 55). In some situations, such as the early development of forest stands on river floodplains in Alaska, browsing of willows during winter by snowshoe hares can be one of the most important factors limiting growth of the plant and influencing stand succession. Likewise, in Scandinavia the grazing pressure on willows and birches during winter by the mountain hare is often reported to have a major impact on the development of the plant community. This pressure obviously poses a serious problem to the plant, but it may have a long-term impact on the browsers as well.

Plants that survive heavy winter browsing often show distinct changes in growth form, as internal allocations of nutrients and carbohydrates are shifted about. For example, birch saplings clipped by browsing moose may respond by producing heavier shoots with larger, more chlorophyll-rich leaves that are retained longer by the plant. In addition, the resting buds on mature branches often produce "long shoots"—shoots with elongated internodes between leaves.[1] Willow responds to heavy browsing by producing numerous stump sprouts of juvenile form, consisting mostly of long shoots without lateral branches.[2] This "compensatory growth" is typical of the response of many species to herbivory and can itself alter the microclimate of the plant community, affecting the future photo-

Fig. 55. Browsing pressure on plants in winter. When the snow is deep and food scarce, plants above the snowpack are exposed to yet another winter stress. In areas where herbivores are abundant, constant browsing pressure on plants may constitute as important a force in natural selection as the worst of winter's weather. As a result, many plants have evolved chemical defenses against browsers, producing distasteful or toxic substances that deter animals from feeding on them. But the browsers need energy, so they coevolve a certain tolerance for the deterrents and keep on eating. Thus goes the struggle between plant and herbivore. Winter is hard on both. (Photos by Peter Marchand.)

synthetic efficiency of the plant as well as influencing later availability of browse to the grazers. Compensatory growth, though, is not the plant's only answer to winter browsing pressure.

PLANT DETERRENTS TO WINTER BROWSING

Several studies of feeding behavior in hares during winter indicate a strong selectivity in browsing preference (see box on "The Eclectic Tastes of Winter Hares"). The green foliage of many (though not all) conifer saplings and evergreen shrubs such as labrador tea are usually passed up in favor of the twigs of deciduous trees and shrubs. Within this preferred group of plants, hares show a consistent pref-

THE ECLECTIC TASTES OF WINTER HARES

Why do hares eat what they eat in winter? There may be any number of factors influencing feeding behavior, not the least of which might be the energy demands of obtaining a particular food item and the perceived danger of exposure while feeding. However, given equal availability of browse materials, either in the wild or in captivity, hares often show a decided preference for one species over another. In such cases, plant selection may be influenced by certain needs on the part of the hare (the "nutritional wisdom" hypothesis) or, alternatively, may reflect strong distaste for specific chemical feeding deterrents produced by the nonpreferred species ("plant defense" hypothesis).

Part of the difficulty in sorting out these hypotheses lies in the variability of food selection shown by hares between one region and another, even where the same species are available. The only constant seems to be avoidance of juvenile plant parts. In any event, this is how plant species from different areas rank among discriminating hares in winter (in descending order of preference):

Alaska and Finland[a]

1. Willow, *Salix* spp.
2. Aspen, *Populus tremuloides*
3. Larch, *Larix* spp.
4. Dwarf birch, *Betula glandulosa*
5. Tree birch, *Betula resinifera*
6. Pine: lodgepole, *Pinus contorta*
   jack, *P. banksiana*
   scots, *P. sylvestris*
7. Fir, *Abies* spp.
8. White spruce, *Picea glauca*
9. Black spruce, *Picea mariana*
10. Alder, *Alnus* spp.

Alaska[b]

1. Paper birch, *Betula resinifera*
2. Aspen, *Populus tremuloides*

3. Balsam poplar, *P. balsamifera*
4. Black spruce, *Picea mariana*
5. Green alder, *Alnus viridis*

Yukon[c]

1. Aspen, *Populus tremuloides*
2. White spruce, *Picea glauca*
3. Dwarf birch, *Betula glandulosa*
4. Willow, *Salix glauca*
5. Balsam poplar, *P. balsamifera*
6. Willow, *Salix alaxensis*

Colorado Rockies[d]

1. Lodgepole pine, *Pinus contorta*
2. Common juniper, *Juniperus communis*
3. Dwarf birch, *Betula glandulosa*
4. Englemann spruce, *P. englemanii*
5. Aspen, *Populus tremuloides*
6. Subalpine fir, *Abies lasiocarpa*
7. Willow, *Salix* spp.

[a]J. P. Bryant and P. J. Kuropat, "Selection of Winter Forage by Subarctic Browsing Vertebrates: The Role of Plant Chemistry," *Annual Review of Ecology and Systemmatics* 11 (1980): 261–85.
[b]J. P. Bryant, F. S. Chapin III, and D. R. Klein, "Carbon/Nutrient Balance of Boreal Plants in Relation to Vertebrate Herbivory," *Oikos* 40 (1983): 357–68.
[c]A. R. E. Sinclair and J. N. M. Smith, "Do Plant Secondary Compounds Determine Feeding Preferences of Snowshoe Hares?" *Oecologia* 61 (1984): 403–10.
[d]P. J. Marchand, unpublished data.

erence for mature growth rather than juvenile shoots, even though the latter may be more plentiful and more accessible during the winter. Evidence suggests that this choice is not always based upon nutrient content or protein level of the preferred forage, but rather may be related to the presence of feeding deterrents in the juvenile shoots and in the buds and catkins of the mature plant. Many plant species, it turns out, have evolved strong chemical defenses in response to grazing pressure—unpalatable or toxic substances that

Table 7. Plant compounds that influence feeding and reproductive behavior of winter herbivores

Plant compound	*Nature in plant*	*Action on herbivore*
Gibberellic acid	Hormone present in highest levels during seed maturation and germination	Doubled number of mice producing litters in lab;[a] possibly important in winter breeding
Tannins (soluble phenolics)	Defense present in flowers and in leaves of forbs, trees, and shrubs	Precipitates plant proteins and GI system enzymes; reduces protein availability and, in some ruminants, cell-wall digestion[b]
Pinosylvin methyl ether (PME)	Defense (toxic phenol) present in buds and catkins of green alder	Repellent to snowshoe hares, mechanism probably as above[c]
Phenolic glycosides	Toxic phenols in willow	Repellent to snowshoe hares[d]
6-MBOA (Methoxybenz-oxazolinone)	Glycoside derivative in vegetatively growing young plants	Accelerates sexual maturity and breeding in voles[e]
Camphor	Toxic phenol in white spruce	Deters snowshoe hare browsing[f]
Papyriferic acid	Toxic phenol in birch	Deters snowshoe hare browsing[g]
3-0 Malonylbetula-folientriol oxide	Toxic phenol in birch	Deters snowshoe hare browsing[h]
Trihydroxydihydro-chalcone	Defense compound in balsam poplar	Deters snowshoe hare browsing[i]

[a]P. Olsen, "The Stimulating Effect of a Phytohormone, Gibberellic Acid, on Reproduction of *Mus musculus*," *Australian Wildlife Research* 8 (1981): 321–25.
[b]C. T. Robbins, S. Mole, A. E. Hagerman, and T. A. Hanley, "Role of Tannins in Defending Plants against Ruminants: Reduction in Dry Matter Digestion?" *Ecology* 68 (1987): 1606–15.
[c]J. P. Bryant, G. D. Wieland, P. B. Reichardt, V. E. Lewis, and M. C. McCarthy, "Pinosylvin Methyl Ether Deters Snowshoe Hare Feeding on Green Alder," *Science* 222 (1983): 1023–25.
[d]J. Tahvanainen, E. Helle, R. Julkunen-Tiitto, and A. Lavola, "Phenolic Compounds of Willow Bark as Deterrents against Feeding by Mountain Hare," *Oecologia* 65 (1985): 319–23.
[e]E. H. Sanders, P. D. Gardner, P. J. Berger, and N. C. Negus, "6-Methoxybenzoxazolinone: A Plant Derivative That Stimulates Reproduction in *Microtus montanus*," *Science* 214 (1981): 67–69.
[f]A. R. E. Sinclair, M. K. Jogia, and R. J. Andersen, "Camphor from Juvenile White Spruce as an Antifeedant for Snowshoe Hares," *Journal of Chemical Ecology* 14 (1988): 1505.
[g]P. B. Reichardt, J. P. Bryant, T. P. Clausen, and G. D. Wieland, "Defense of Winter-Dormant Alaska Paper Birch against Snowshoe Hares," *Oecologia* 65 (1984): 58–69.
[h]P. B. Reichardt, T. P. Greene, and S. Chang, "3-0-Malonylbetulafolientriol Oxide I from *Betula nana* subsp. *exilis*," *Phytochemistry* 26 (1987): 855–56.
[i]M. K. Jogia, A. R. E. Sinclair, and R. J. Andersen, "An Antifeedant in Balsam Poplar Inhibits Browsing by Snowshoe Hares," *Oecologia* 79 (1989): 189–92.

Fig. 56. Mature (left) and juvenile twigs of Alaskan paper birch. Chemical defense of young shoots against browsers like the snowshoe hare is so pronounced in this species that resin droplets of the deterrent chemical actually accumulate on the surface of the juvenile twig.

serve as deterrents to browsing. These, along with other plant compounds that may affect winter browsers, are summarized in table 7. Plants that typically inhabit high-resource environments and have rapid growth rates, such that they soon grow out of the reach of grazers, may produce antiherbivory compounds only in their juvenile parts. Other plants inhabiting low-resource environments and having slow growth rates with less potential for compensatory growth typically maintain chemical defenses throughout their life cycles.[3]

In Alaskan paper birch, the chemical compound that apparently serves as the deterrent is papyriferic acid, which is present in juvenile stems at concentrations 25 times greater than in mature stems.[4] So great is the defense level of this species in boreal forest regions of Alaska, where browsing pressure is especially high, that resin droplets of the chemical actually accumulate on the surface of juvenile twigs (fig. 56). The effectiveness of papyriferic acid as a feeding deterrent has been demonstrated by treating oatmeal with purified extracts from the birch resins and offering it, along with untreated oatmeal, to captive snowshoe hares. In these experiments the hares consumed only one-fourth as much of the treated oatmeal as the untreated feed, and when offered only juvenile birch twigs with naturally high concentrations of the acid, the captive hares eventually stopped eating entirely.[5]

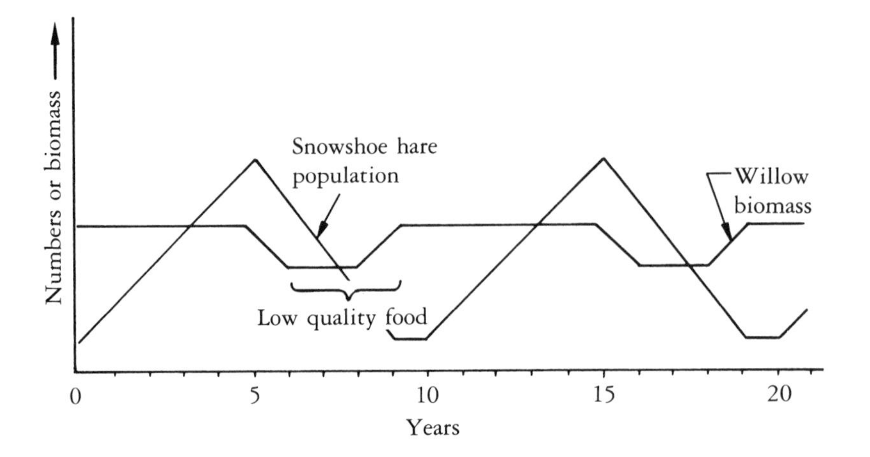

Fig. 57. Time lag in recovery of hare population keyed to plant defenses. Willows and birches respond to overbrowsing by producing juvenile shoots with strong chemical defense compounds that deter further browsing. These chemical defenses are maintained for two to three years, thus creating a food shortage that extends beyond the recovery of plant biomass following the hare population decline. This delay in recovery of food quality probably contributes to the lengthened periodicity of snowshoe hare population cycles. (Adapted from L. B. Keith, "Role of Food in Hare Population Cycles," Oikos, *40 (1983): 385–95.)*

Avoidance of juvenile willow branches by hares may be related to higher lignin content and lower digestibility of nitrogen[6] or to increased concentrations of the compound phenolic glycoside.[7] When free-ranging mountain hares in Finland were offered bundles of different-aged willow and aspen twigs along with oats that were either untreated or treated with extracts of phenolic glycoside from willow bark, the results were similar to that seen with the birch. Mountain hares showed a strong preference for mature shoots of both species and avoided the treated oats in favor of the untreated, apparently able to discriminate between the food items by scent alone.[8]

Some researchers suggest that preferential browsing of mature willow shoots by snowshoe hares may have an important effect on the hare population itself. In boreal forest regions of North America and the Soviet Union, hare population cycles have a regular periodicity of 8 to 11 years, which is thought by some to be the result of a time delay in vegetation recovery following overbrowsing during winter. As we have seen here, when some woody plant species are browsed heavily, they respond by producing juvenile shoots that are of very low food value and that typically maintain their chemical defenses for a period of two to three years. In effect, as browsing pressure increases during the growth phase of the hare population cycle, the resistance of plants to further browsing also increases. This, then, may be one mechanism that limits hare population growth and initiates the declining phase of the population cycle.[9] The evolved chemical response of plants to browsing effectively extends the food shortage beyond that induced by overbrowsing alone and, thus, provides a time lag in recovery of food resources (fig. 57). Meanwhile, declining hare reproduction (induced by a scarcity of high-nutrient food resources) and high winter mortality, perhaps coupled with increased predation pressure following the population peak, would continue to drive the hare population down.

The notion that food quality is particularly important during winter, and that browsers have the ability to select food of high nutrient content, is inherently appealing to our sense of reason. The notion that plants under particularly heavy winter browsing pressure must evolve some means of countering such pressure (or be browsed out of existence) is also intuitively appealing. The problem, however, is that the situation in the field is often extremely complex, with both browsers and plants continually responding to everchanging needs, and the data do not always agree with our sense of what ought to be.

As a case in point, the hypothesis depicted in figure 57 predicts that plant defense compounds should be low during the bottom phase of a hare population cycle and high during and just after the peak of the hare cycle. But researchers at the University of British Columbia, led by A. R. E. Sinclair, have been following trends in food quality, hare feeding preferences, and population cycles in the Kluane Lake District of the Yukon for over 10 years and find no such correlation.[10] In fact, they report, contrary to studies so far cited, food selection was not related to the presence of chemical deterrents. Instead, they discovered that both captive and wild hares ate large amounts of plant material containing allegedly defensive compounds, and they generally consumed different foods in amounts roughly equal to their availability. While the hares showed a decided preference for mature rather than juvenile plant parts, regardless of species, some of the mature twigs had higher resin concentrations than younger plant parts.[11] Similarly, favored winter foods of snowshoe hares in the central Rockies are browsed in proportion to their availability, and many of the more commonly eaten plant species often contain significantly higher concentrations of allegedly unpalatable resins than the less common food items. Consistent rejection of juvenile plant parts by hares may, therefore, be related to something other than the presence of plant-defense compounds.

COEVOLUTION OF PLANTS AND BROWSERS

It is quite possible that the effectiveness of plant secondary compounds as feeding deterrents is related to the coevolutionary history of plant and consumer, particularly with regard to the intensity of winter-browsing activity and length of time involved. In a very ambitious and well-conceived experiment to test this idea, John Bryant and his colleagues at the University of Alaska in Fairbanks recently collaborated with Jorma Tahvanainen and co-workers in Finland to survey, on a circumpolar scale, browsing preferences using plants and hares of similar ecological affinity but differing in coevolutionary history.[12] These researchers hypothesized that plants from the oceanic island of Iceland, which had no browsing animals before Norse colonization, would produce fewer chemical feeding deterrents during winter than would closely related species in Finland, where hares have long been abundant but do not exhibit the pronounced 10-year population cycles observed in mainland Siberia and Alaska. Plants from the latter, they surmised, would produce the highest levels of defense compounds due to exposure over a long

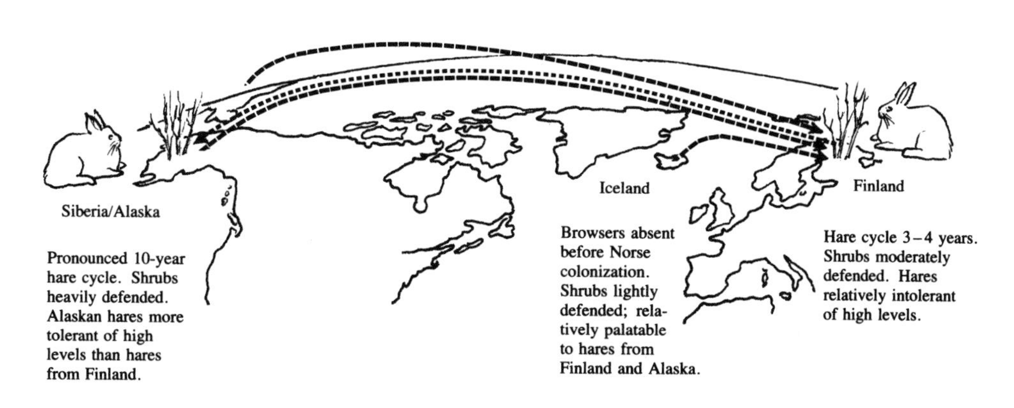

Fig. 58. Summary of experiment showing the relationship between evolutionary history of plant-browser interaction, levels of chemical defense in the plant, and tolerance of feeding deterrents by hares in winter. (From data of J. P. Bryant, J. Tahvanainen, M. Sulkinoja, R. Julkunen-Tiitto, P. Reichardt, and T. Green, "Biogeographic Evidence for the Evolution of Chemical Defense by Boreal Birch and Willow against Mammalian Browsing," American Naturalist *134 (1989): 20–34.)*

period of time to particularly heavy browsing pressure, especially as hare numbers peak every decade. On this premise they predicted that hares from all regions, if given a choice, would preferentially browse first Icelandic plants, then Finnish plants, and lastly Siberian and Alaskan plants. They further predicted that Alaskan snowshoe hares, because of a longer and more intensive coevolution with the browse species, would show a greater tolerance for chemical deterrents than would Finnish mountain hares.

To test their hypotheses, samples of birch and willow from different regions were collected in the field and flown back and forth between Finland and Alaska to be offered to both free-ranging and captive hares. At the same time, feeding trials were conducted in previously established garden plots in Finland containing birches from Iceland, Finland, and Siberia. In short, the reciprocal experiments of these researchers confirmed each of their predictions. Chemical feeding deterrents were lowest in Icelandic birch, intermediate in Finnish plants, and highest in the Alaskan and Siberian shrubs; and hares preferred the various food sources in the same order (fig. 58, p. 173). Also, while both Alaskan and Finnish hares would eat Finnish birch and willow before Alaskan shrubs, the captive Alaskan hares would eat Alaskan willows containing high concentrations of phenolic glycosides when Finnish hares refused them, suggesting that Alaskan hares had indeed evolved a greater tolerance for the feeding deterrents.[13]

In all studies to date, extracts of plant phenols and resins have been shown to reduce palatability of food to hares when applied experimentally to preferred foods like oatmeal or commercial rabbit chow. In spite of this, we have seen that free-ranging snowshoe hares during winter sometimes select food with high concentrations of resins. These plants, it turns out, often have a higher digestibility than less-preferred foods with a much lower resin content.[14] Thus, it appears that other needs on the part of the browser may prevail, with nutritional gains from eating a heavily defended species sometimes outweighing the unpleasant consequences. And nutritional needs may be everchanging, constantly affecting browse preferences.

PLANTS AND THE QUALITY OF SUBNIVEAN LIFE

Fluctuations in microtine rodent populations in Finland have also been linked to plant-animal interactions, in this case involving changes in both plant nutritive quality and chemical responses to grazing.[15] Arctic and subarctic plants themselves exhibit cyclic patterns in flowering and fruit production due to the general scarcity

Fig. 59. Cyclic patterns of plant production and rodent populations in northern Finland (From S. Eurola, H. Kyllonen, and K. Laine, "Plant Production and Its Relation to Climatic Conditions and Small Rodent Density in Kilpisjarvi Region (69°05'N, 20°40'E), Finnish Lapland." in Winter Ecology of Small Mammals, *ed. J. F. Merritt, Carnegie Museum of Natural History Spec. Publ. 10 (Pittsburgh, 1984), 121–30.)*

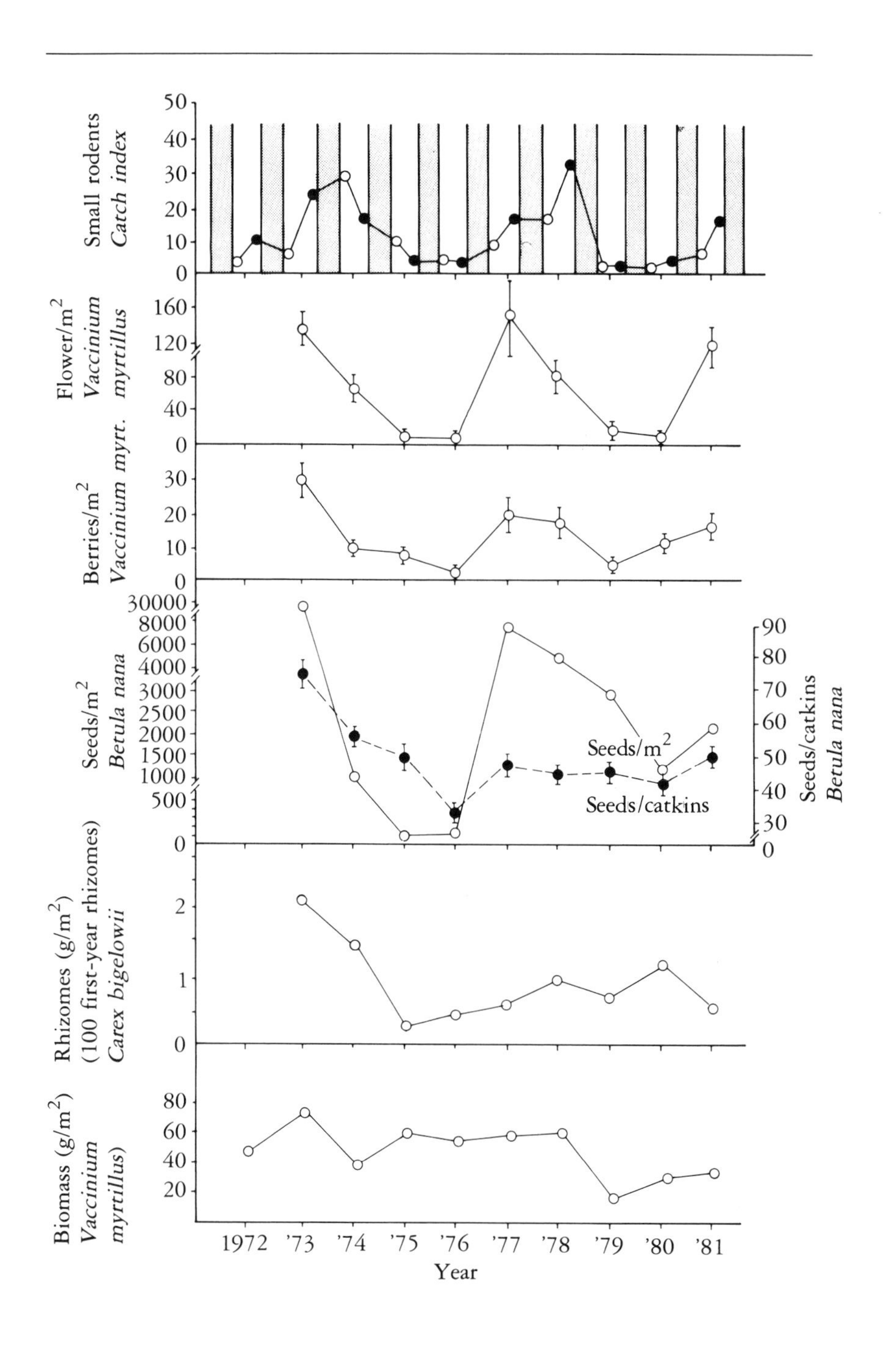
Small rodents
Catch index
Flower/m²
Vaccinium myrtillus
Berries/m²
Vaccinium myrt.
Seeds/m²
Betula nana
Seeds/m²
Seeds/catkins
Seeds/catkins
Betula nana
Rhizomes (g/m²)
(100 first-year rhizomes)
Carex bigelowii
Biomass (g/m²)
Vaccinium myrtillus)
1972 '73 '74 '75 '76 '77 '78 '79 '80 '81
Year

of resources in that environment. One growing season is rarely long enough to allow accumulation of the necessary carbohydrate and nutrient reserves for flowering, so plants accumulate reserves for several summers until some threshold is reached and flowering commences. It follows, then, that plants in flowering condition are more vigorous and of potentially greater food value. During the winter prior to flowering, rootstocks or rhizomes of perennial herbs are high in carbohydrate reserves and provide an important source of energy to subnivean mammals. During the winter following flowering, high-energy seeds or fruits are abundant. By the second winter after flowering, the nutritive value of the plant is low again. Cyclic patterns of plant growth are not always synchronous among species in northern Finland, but the periodicity of flowering and seed production is fairly regular at three- to four-year intervals and corresponds closely with the rise in small mammal populations, often preceding the peak by one year (fig. 59, p. 175).[16]

Following heavy grazing by small mammals, these herbaceous plants may also increase production of defense compounds, thereby having additional impact on the herbivore populations. Because synthesis of defense compounds competes with other growth functions, such as leaf production, for nutrients and metabolites, an increase in production of antiherbivory compounds in areas where nutrient resources are scarce is likely to slow the rate of plant regrowth. Recent observations in Finland suggest that induced chemical response by herbaceous and woody plants rises and falls with changes in grazing pressures. The ratio of toxic phenols to nitrogen, for example, in the tundra bilberry (*Vaccinium myrtillus*) nearly doubled in the two-year peak and decline of a lemming population cycle.[17] This may have the same time-delay effect on animal population recovery as suggested in the snowshoe hare birch–willow interaction.[18] Superimposed on the relationships described here are the effects of increased predation following the peak in rodent numbers, and possible parasite and disease outbreaks which also affect the rate of population recovery.[19] These interactions are summarized in figure 60. It is interesting to note that arctic hare populations in northern

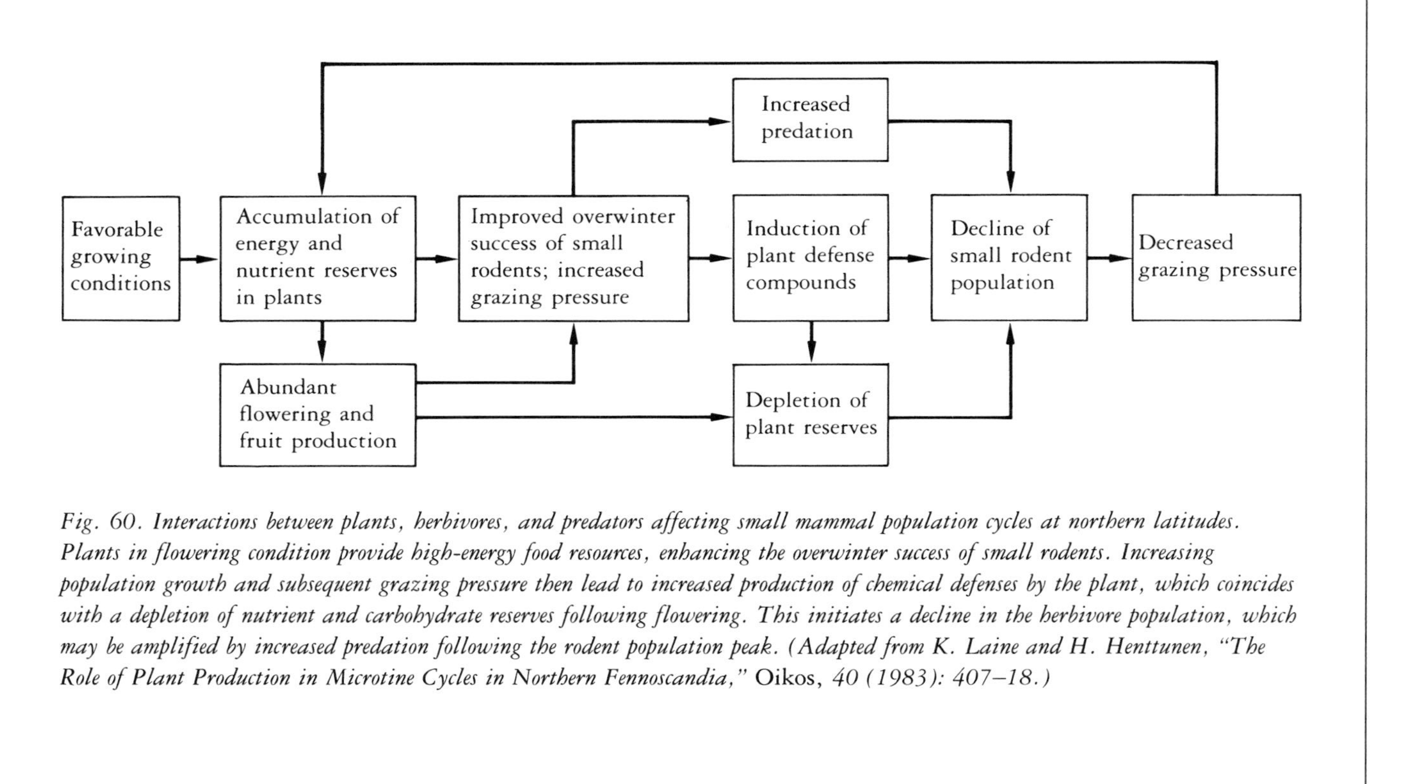

Fig. 60. Interactions between plants, herbivores, and predators affecting small mammal population cycles at northern latitudes. Plants in flowering condition provide high-energy food resources, enhancing the overwinter success of small rodents. Increasing population growth and subsequent grazing pressure then lead to increased production of chemical defenses by the plant, which coincides with a depletion of nutrient and carbohydrate reserves following flowering. This initiates a decline in the herbivore population, which may be amplified by increased predation following the rodent population peak. (Adapted from K. Laine and H. Henttunen, "The Role of Plant Production in Microtine Cycles in Northern Fennoscandia," Oikos, *40 (1983): 407–18.)*

Scandinavia follow the same three- to four-year cycle of microtine rodents rather than the 10-year cycle of snowshoe hares in North America and the Soviet Union. This is apparently because predators switch over to the hares during the decline of rodent populations.[20]

We should not overlook the possibility, too, that plants might directly mediate the occasional winter breeding that is documented for some subnivean mammals. Increasing evidence suggests that the stimulus for breeding in small mammals may come from plant compounds ingested by the animal. Compounds that induce reproductive activity in captive rodents have been isolated and include gibberellic acid, a plant growth hormone that is most abundant during germination and seed maturation.[21] In flowering years when ripening seeds are available, increased gibberellic acid ingestion could amplify the effects of improved food quality on population growth by stimulating breeding before winter. Germination of that same seed crop under snow, which is known to be possible for a number of plant species,[22] might also stimulate breeding during winter in the subnivean environment, though insufficient data presently exist on the periodicity of winter breeding to link it conclusively with plant growth cycles. Nevertheless, winter breeding of the root vole in northern Finland has been known to occur only when young shoots of cotton grass were found just prior to winter with leaves newly opened.[23]

THE CARBON DIOXIDE DEBATE

Finally, we come to an interaction of an indirect nature—and a question of considerable controversy in the literature of winter ecology—involving the role of plants and soil microbes in carbon dioxide buildup under snow and its possible effects on subnivean animal activity. The respiration of plants buried under snow, including the root systems of large trees as well as the small perennial herbs of the ground layer, coupled with the activity of decay organisms such as the abundant heterotrophic fungi that break down the dead organic material of the forest floor, results in the liberation of CO_2 under the snowpack. In some cases, the release of CO_2 during winter has been found to be negligible,[24] but in other cases it may be considerable. Even though plant and microbial contributions are greatly reduced at low temperatures, soil respiration as high as 50 cc of CO_2 per hour for a square meter of ground surface has been measured during winter.[25] If the snowpack were to act as a barrier to the diffusion of CO_2 in those instances where its emission is substantial, then it might be expected that animals active in the subnivean environment would suffer some deleterious effects from CO_2 buildup.

In 1946 the Soviet ecologist A. N. Formozov wrote of vertical tunnels constructed in the snowpack by voles that "probably serve to ventilate the deeper parts of the burrow."[26] He also observed that "after each fresh snowfall the voles clear these 'windows,' but very rarely come out of them." The notion that subnivean mammals deliberately construct ventilation shafts to diffuse CO_2 accumulating under the snowpack has persisted in the folklore of winter ecology since that time. But some of the observations relating to the use of these tunnels are contradictory, and the few attempts at measuring subnivean CO_2 levels during winter have produced inconsistent results. By one account, ventilation shafts are used only in the early part of winter and become uncommon after mid winter.[27] If their use is related to CO_2 accumulation, the reasons for their disappearance after mid winter is perplexing. At the Kilpisjarvi Biological Station in Finland, K. Korhonen counted anywhere from 8 to 18 "air vents" per 100 m^2 of snow surface—an unusually high number—but never found CO_2 levels under the snow high enough to suspect any physiological effect on the voles.[28] Contrary to Formozov's observation, he noted, too, that voles showed no haste in reopening holes that had been filled with snow, and that many areas with large vole populations had no vents at all. Korhonen also reported that CO_2 diffused very readily in snow, even under wet or highly compacted snow.

On the other hand, several researchers have recorded subnivean CO_2 accumulations substantially greater than ambient levels,[29] one of the highest concentrations ever reported approaching 4% under

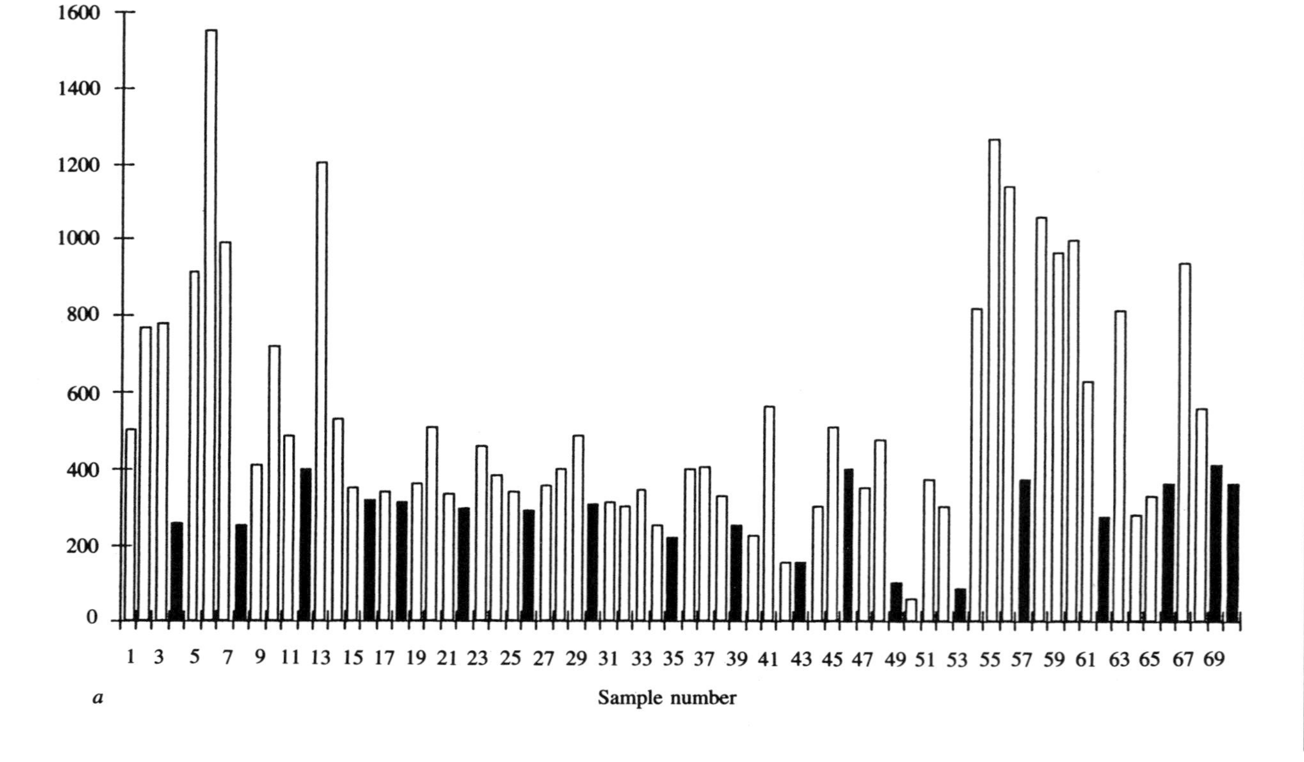

CO2 in ppm
1600
1400
1200
1000
800
600
400
200
0
1 3 5 7 9 11 13 15 17 19 21 23 25 27 29 31 33 35 37 39 41 43 45 47 49 51 53 55 57 59 61 63 65 67 69
Sample number
a

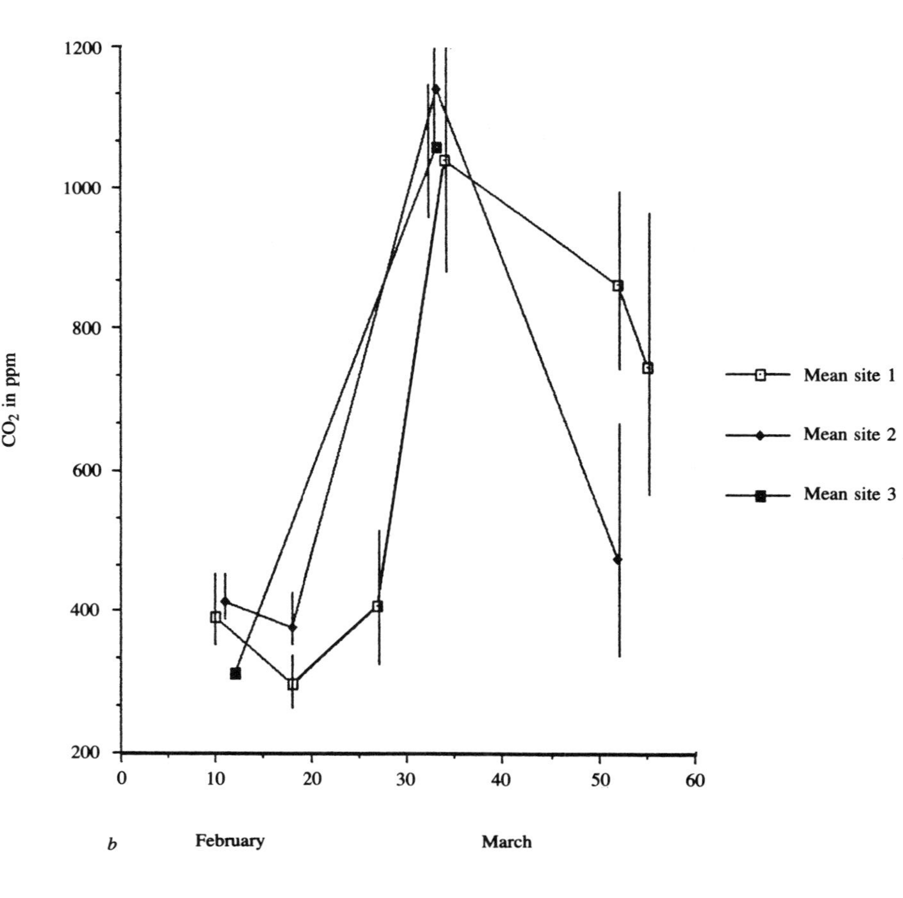

Fig. 61. Accumulation of CO_2 under the snow. CO_2 levels (a) *under the snowpack (unshaded bars) often show dramatic departures from atmospheric levels immediately above the snow (shaded bars). The first 22 samples here are from a number of sites in the Colorado Rockies and the remaining from abandoned field and woodland sites in Vermont. Vegetative character, snowpack conditions, and weather all influence subnivean accumulations of CO_2, though sites near each other tend to fluctuate in like manner* (b)*, regardless of vegetation differences. (From M. C. Young, P. J. Marchand, and C. A. Bryant, "Carbon Dioxide Accumulation beneath Snow and the Response of Voles to Elevated CO_2 under Simulated Subnivean Conditions," unpublished.*

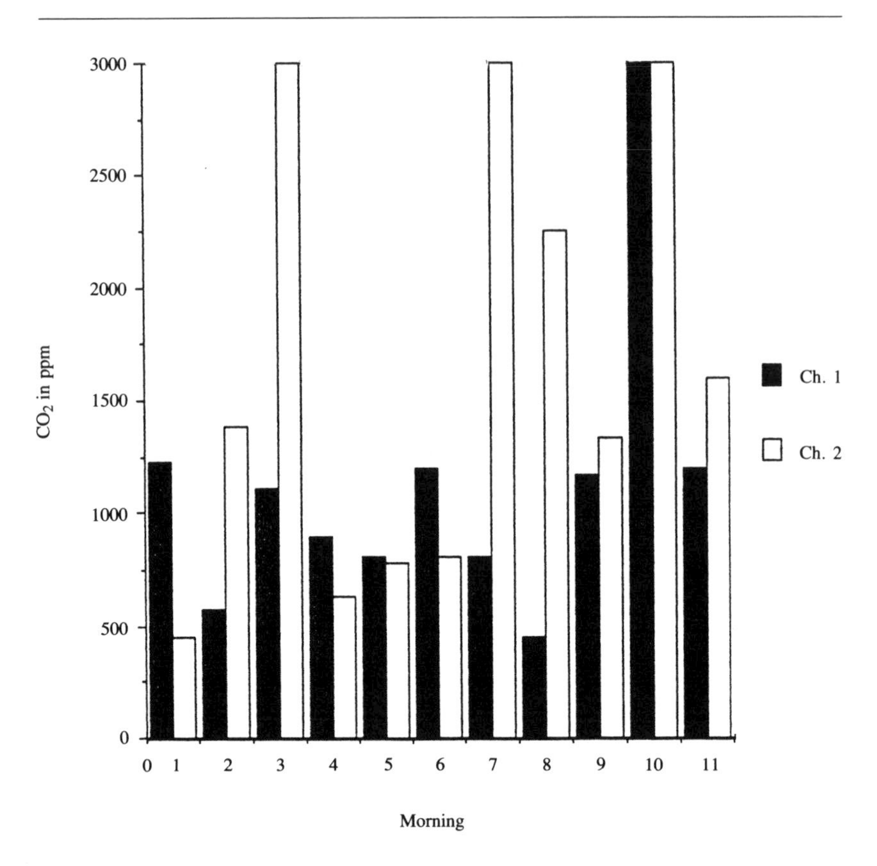

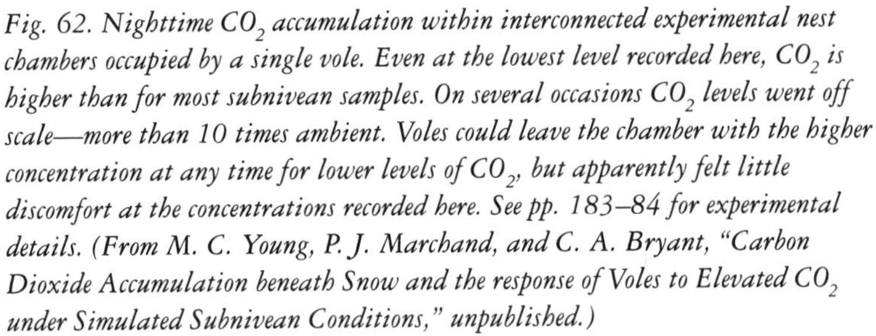
Fig. 62. Nighttime CO_2 accumulation within interconnected experimental nest chambers occupied by a single vole. Even at the lowest level recorded here, CO_2 is higher than for most subnivean samples. On several occasions CO_2 levels went off scale—more than 10 times ambient. Voles could leave the chamber with the higher concentration at any time for lower levels of CO_2, but apparently felt little discomfort at the concentrations recorded here. See pp. 183–84 for experimental details. (From M. C. Young, P. J. Marchand, and C. A. Bryant, "Carbon Dioxide Accumulation beneath Snow and the response of Voles to Elevated CO_2 under Simulated Subnivean Conditions," unpublished.)

snow in areas of low relief near Moscow (ambient levels of CO_2 are generally close to 0.03%).[30] The majority of these reports show subnivean CO_2 levels between 3 and 10 times ambient. One study compared subnivean CO_2 concentrations in an area having no small mammals to that within a 35 m^2 caged area containing 10 field voles. The voles' enclosure was found to have on average three times more CO_2 and much greater variability than the non-inhabited area, suggesting that mammalian contributions to subnivean accumulation might itself be significant.[31]

The larger question, however, is whether or not the CO_2 levels commonly seen under the snowpack can be considered deleterious in any way to nonhibernating subnivean mammals. Cheryl Penny and William Pruitt at the University of Manitoba, in a unique field study, reported small mammal movement out of winter habitats in which CO_2 buildup consistently occurred.[32] Experimentally, however, it is difficult to demonstrate that the levels typically (or even exceptionally) measured under the snowpack cause problems for voles. Matthew Young, in my own laboratory, spent considerable time characterizing the subnivean atmosphere for different habitats in Colorado and Vermont, and then devised an elaborate experimental scheme for monitoring behavioral response of voles to elevated CO_2.[33] He exposed captive animals to artificially high levels while offering them an opportunity to escape to normal atmospheric levels at any time. His experimental apparatus consisted of two identically provisioned nest chambers connected by a tunnel with a treadle-operated door through which a captive vole could move with complete liberty (which it did frequently). The whole apparatus was placed in a controlled-environment chamber to simulate as closely as possible subnivean light and temperature conditions, and was fitted with tubing connected to a CO_2 source and an infrared gas analyzer for continuous flow-through air monitoring. Winter-acclimatized voles were allowed a period of time to "settle in" to one chamber or the other, and then the CO_2 level was elevated in whichever chamber the vole happened to be in at the start of a trial. The unoccupied chamber, now its "escape" chamber, was maintained at the ambient CO_2 level. The behavioral response of the vole was then observed carefully.

The only conclusion that could be drawn from this study is that, while high subnivean CO_2 levels were confirmed on occasion—varying greatly with habitat, time, weather (especially wind), and snowpack conditions (fig. 61, pp. 180–81)—voles under experimental conditions generally appear oblivious to CO_2 levels as much as 10 times ambient. The most convincing evidence that they are unaffected physiologically by such high levels came somewhat unexpectedly, when it was discovered in turning on the gas analyzer in the morning, that overnight CO_2 levels in the vole's sleeping chamber sometimes went off scale just from normal respiration—over 3000 ppm or more than 10 times ambient—and this with the tunnel door locked open and the chamber continuously purged by a slow through-flow of fresh air (fig. 62, p. 182)! Thus, it has been difficult to find convincing support of claims that vertical tunnels to the surface of the snowpack represent deliberate attempts on the part of small mammals to ventilate the subnivean space. Given the risks, however, of extrapolating from carefully controlled laboratory experiments to the infinitely more complex field situation, it might be prudent, until we can come up with more field data on vole behavior, to leave these questions open: Do CO_2 levels under the snow at least occasionally reach deleterious levels and, if so, under what conditions (of weather, habitat, snowcover, plant activity)? How do subnivean mammals respond physiologically or behaviorally to such stresses if they do occur? And what role might this play in the overwintering success of these animals?

7 WINTER PROFILES: A SEASON IN THE LIVES OF SELECTED ANIMALS

It should be apparent by now that there are few absolutes in the science of winter ecology: that many of the "rules" for overwintering based on physical or chemical energetics have been compromised by animals that are yet undeniably successful in snow country. We have seen, for example, that while large body size confers certain thermal advantages in cold climates, many animals systematically reduce body mass in preparation for winter (pp. 118–20). We concluded that white coloration may confer insulative benefits to an animal, in addition to offering protective coloration (pp. 121–25), but only a small number of mid-sized animals take advantage of either by molting to white for winter. We saw that freezing within animal tissues may cause a number of deleterious effects (pp. 127–28), but that many insects, as well as painted turtle hatchlings and possibly other reptiles, deliberately seed ice crystals in their bodies to promote freezing at temperatures only a little below 0° C (pp. 133–35). In the process of natural selection, the energetic advantages of some attributes have been traded for the practical gains of others. "Laws" have been violated for the sake of more efficient predation (p. 118) or for the benefits of reduced competition among nestmates (p. 121).

Overwintering success, we see, depends not upon the perfect solution to low temperatures or deep snowcover, but upon the entire suite of adaptations by which animals are able to profitably exploit all of the resources of their environment, while at the same time maximizing their ability to avoid predation. In this chapter I would like to look at a few specific animals to see how behavior and physiology combine to mitigate the winter stresses particular to various "occupational" specializations. Three groups of animals will serve to illustrate a number of different overwintering strategies.

THE NORTHERN CERVIDS (FROM CERVIDAE, THE DEER FAMILY)

Deer are often among the more visible of animals active during winter in snow country—visible both by their extraordinary versatility and success in occupying diverse habitats, often in close association with humans, and visible by their occasional failure during severe winters, congregating in numbers and dying of malnutrition for lack of adaptation to deep snow. Their cold tolerance is nonetheless rather remarkable, and, excepting the severest of conditions, the more northern members of this family, moose and caribou, may actually experience less environmental stress in winter than they do with the heat and biting insects of summer.

Paradoxically, both the success and failure of cervids in the North has much to do with their being ruminants. It is the ability of deer in winter to assimilate woody plant material through microbial fermentation that is both their advantage and their limitation. As ruminants, they are assured of something to eat almost anywhere, provided only that they can reach it. But browsing woody plants does not yield equal benefit at all times of the year. In winter the nutritional value of deciduous trees and shrubs is diminished, effectively diluted by a greater percentage of structural "packaging" in the form of thickened and lignified cell walls. Thus, in contrast to the abundant protein and soluble carbohydrates of fresh leaves, the bulk of winter browse is cellulose, which takes longer to break down in the rumen of the animal.

To some extent, this dietary constraint is balanced by increased heat production associated with the extended digestion process in ruminants. An animal sitting quietly at rest, but ruminating, exhibits a metabolic rate higher than its basal metabolism by an amount associated both with the activity of ruminating and with the heat of fermentation arising from microbial action in the rumen. This heat increment of feeding, which also includes energy expended by digestive tissues, as well as heat from increased metabolism of tissues outside of the gastrointestinal tract[1] (sometimes referred to as the "specific dynamic action of food"), may contribute importantly to thermoregulation and can significantly depress the lower critical temperature of a ruminant.[2] The amount of heat so produced may range from 10 to 50% higher than the fasting heat production of wild cervids (though the higher rate is probably not sustained for long).[3] Still, the amount of food a ruminant can process is constrained by how long it takes to digest it, and for this reason, the rate of energy intake in winter is usually limited, even with plenty of browse material available (this is why deer have been known to die of malnutrition with their stomachs full). To cope with this limitation during times of high energy demand, all cervids in winter rely heavily on fat reserves accumulated earlier in the year.

With regard to fat storage, the cervids benefit from a size advantage that we have not discussed previously. The capacity of an animal to store fat is directly related to its mass (i.e., capacity and size vary in a one-to-one proportion), while the rate at which it utilizes energy to maintain body temperature is proportional to some fractional power of its size, usually three-fourths—a relationship that seems to hold for a great number of mammals. The difference in these two properties means that as body size increases, the ability to store fat increases faster than the rate at which fat is metabolized (fig. 63). Thus, while a small nonhibernating mammal must eat more frequently, having a narrower margin between fat storage capacity and rate of consumption, large animals can remain warm (assuming sufficient metabolizable fat stores) for considerably longer periods without eating.[4] Whether an animal will store fat or not is species dependent and often related to environmental characteristics such as seasonal food availability, but clearly a large animal is less constrained physically than a small one in terms of the amount of fat it can add before mobility is impaired.

In wintertime, then, when food is relatively scarce or of poor quality, the cervids are confronted with two very different strategic possibilities for survival: they may either increase foraging effort to compensate for the lower nutrient availability and higher energy requirements at colder temperatures, or they may reduce foraging effort to conserve energy, relying on supplemental reserves to compensate for lower food intake. While domestic herbivores usually

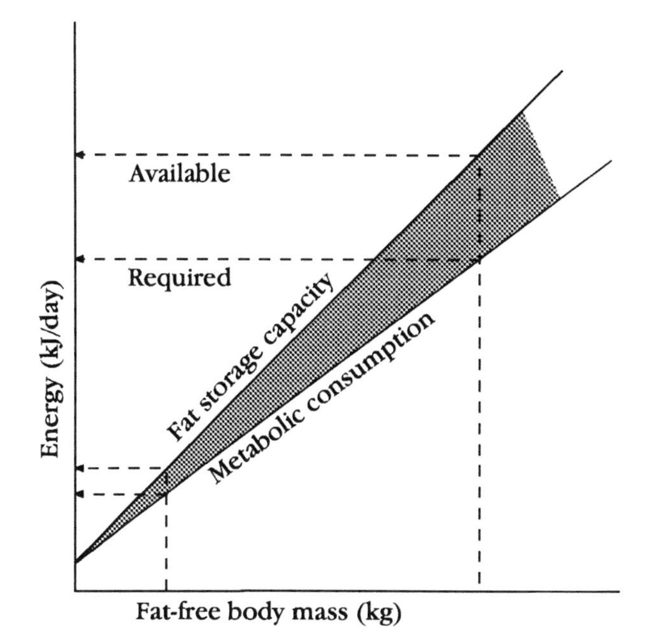

Fig. 63. Schematic relationship between body size, capacity to store fat, and metabolic rate for homeothermic animals. Both coordinates are logarithmic. The shaded area represents maximum surplus energy, indicating that larger animals with adequate fat reserves can subsidize a greater percentage of their energy needs during times of food scarcity. (After A. R. French, "The Pattern of Mammalian Hibernation," American Scientist *76 (1988): 569.)*

respond to scarcity or greater need by increasing effort, wild ruminants often conserve energy instead.[5] Overwinter success thus becomes a challenge of balancing combined heat loss to cold surroundings plus energy expended on foraging effort (as well as other activities) against realized gains from food intake plus energy available from the metabolism of fat reserves stored during times of improved food quality. The northern cervids have evolved a wide range of physiological and behavioral responses to meet these challenges, differing to a degree in their effectiveness and explaining much in terms of species range limits.

---

*White-Tailed Deer (*Odocoileus virginianus*)*
*and Mule Deer (*Odocoileus hemionus*)*

Both white-tailed and mule deer are distributed over an extraordinarily broad geographical range, stretching from southeastern Alaska, across northern British Columbia and Alberta (mule deer) to the southern end of James Bay (white-tail), thence south to the Mexico border and beyond. Along this latitudinal gradient, temperatures may range from as low as −60° C to nearly +50° C.[6] While there is some degree of physiological adaptation exhibited by individuals at the climatic extremes, mostly in adjustments of insulative thickness, metabolic rate, and water requirement, the remarkable success of both species in the North appears equally dependent upon behavioral adjustments to scarce or poor quality food and to the encumbrance of deep snow. Some of this behavior has much to do with being a ruminant, as we discussed earlier, but other behaviors are clearly of advantage in snow, compensating for a lack of specific morphological adaptation.

Neither species is endowed with any special physical ability to cope with snow. Snow depths of 25–30 cm greatly impede mobility of both, and often accumulations of nearly half that will stimulate movement of deer to areas of reduced snowcover. Mule deer in montane forest habitats frequently migrate to lower, more open, south-facing slopes, usually shrub dominated, or to windswept ridges where snow depths are less (fig. 64).[7] In more northern interior portions of their range where deep snow is unavoidable even at lower elevations, mule deer preferentially seek old-growth conifer stands in which the dense canopy intercepts more snow and radiates more long-wave energy back to the ground (see discussions of radiant energy exchange on pp. 25–26 and 96–97). Where mule deer territories include Douglas fir stands, this tree is favored for an additional advantage: compared to other species, the older Douglas firs tend to produce more litterfall of foliage and arboreal lichens, which are preferred by the deer over the foliage of younger trees.[8] Similarly, white-tailed deer in the northern part of their range show a strong tendency

Fig. 64. Lacking specific adaptations for deep snow, mule deer in the western United States often move to drier, shrub-dominated slopes for winter, where browse is generally more accessible. (Photo by Bruce Gill.)

to congregate or "yard" in areas of dense conifer growth where snow cover is less and their collective trampling and trail establishment facilitate somewhat easier movement. Counteracting these advantages, however, such behavioral response to snow may confine mule and white-tailed deer to 10 to 20% of their normal foraging territory,[9] putting considerable pressure on limited food supplies.

Just how critical food shortages might become in the area of deer yards is often difficult to project. The daily energy expenditure of free-ranging deer in winter is influenced by a number of factors, including everchanging temperatures, snowpack conditions, accessibility of browse, and the activity budgets of individual animals. Nonetheless, the early work of Aaron Moen provides a starting point for estimating the caloric demands imposed on deer by low air temperatures and snow of varying depths. Using a simplified energy budget approach, Moen calculated daily metabolic expenditures for different amounts of time spent in various activities such as bedding, standing, and foraging in snow of varying depths. At one end of his activity spectrum, Moen predicted a total energy expenditure of 7400 kJ (divide by 4.184 for kcal) per day for a 60-kg animal bedded 75% of the time and walking slowly 5% of the time in shallow snow (the remainder of the time the deer was either standing or foraging). At the other end of the spectrum, an animal bedding only 25% of the time, with a proportional increase in all other activity levels, showed a predicted energy expenditure of 10,000 kJ. The difference between these two levels of activity amounted to an increased browse requirement of over .6 kg of fresh material per day.[10] At a snow depth of 50 cm, the same increase in activity would require an additional 1.0 kg per day of fresh browse material, a supplemental demand of significant proportion when multiplied by the number of deer typically concentrated in a yard.

Moen's estimates focused attention on the high energetic costs of activity for an animal morphologically unadapted to deep snow—a problem confirmed once it became possible to measure metabolic rates under field conditions. It has since been shown that a white-tailed deer, merely by standing, experiences a 20% increase in metabolic rate over that of a bedded animal, as a result of the greater surface area exposed to heat loss. Foraging activity may increase energy expenditure by 28 to 33% over lying.[11] What Moen underestimated, however, was the dramatic escalation in the energy cost of locomotion for a single animal with deepening snow. Studies with mule deer, elk, and caribou have shown an exponential gain in energy expenditure as both the sinking depth of an animal in the snow and its rate of travel increase (see fig. 66 on p. 200 in following section). Greater snow density has a similar effect, increasing foot drag and energy costs accordingly.[12] Only through behavioral adjustments, namely the repeated use of trails in deep snow, can deer compensate for this expense.

While direct comparisons of metabolic expenditure for different animals are complicated by variations in experimental approach, age of test animals, season, and preconditioning, Moen's estimates probably represent the lower limit of daily energy requirements for undisturbed cervids in winter,[13] particularly at the northern limits of their range. The more important contribution of Moen's seminal work, however, was not in the absolute values of his predictions, but rather in his demonstration in the field that whenever weather conditions favored increased heat loss from animals, free-ranging white-tailed deer decreased their level of activity, as revealed by track analysis. Voluntary reduction of activity, it appeared, was their best means of covering energy deficits, a conclusion since borne out by several other studies. Though deer may expand their food choices fourfold as snow conditions or forage depletion dictate,[14] current evidence suggests that even under the best of circumstances, deer cannot satisfy their daily energy requirements from browse available in winter,[15] and thus draw on fat reserves accumulated earlier in the year to balance their needs. That this response is controlled by some internal mechanism is suggested by additional studies showing that mule deer on a high-quality diet still lose on average 20% of their peak weight during winter.[16] Given adequate fat reserves, then, it appears that for deer, the energy savings incurred through voluntary reduction of activity outweigh potential gains from increased foraging effort.

On the heat loss side of the ledger, both the white-tailed and mule deer fare quite well. In fact, the capacity of deer in northern parts of their range to adjust their lower critical temperature downward for winter is possibly unsurpassed by any other animal. This is accomplished in part through effective constriction of surface blood vessels and peripheral cooling, primarily in the extremities.[17] However, most of the decline in lower critical temperature occurs through seasonal pelage change and the superior winter insulation of the animal's trunk. Deer undergo a fall molt in which a marked increase in the density and length of both guard hairs and underfur occurs. When the winter pelage of northern white-tailed deer reaches its full growth, usually by December, guard hairs approach

5 cm in length with a density of up to 1000 per cm.[2] These are stiff, hollow, and somewhat wavy or crimped lengthwise, and thus provide considerable insulative benefit in addition to protecting the very dense underfur. The fine hairs of the underfur may grow to 2 cm or more in length by early winter[18] (see box on "Fur—The Mammalian Advantage"). Overall, the insulative value of deer pelage in winter is similar per unit depth to that of the caribou,[19] resulting in a lower critical temperature approaching −20° C, nearly 30° below that of summer.[20] Thermoregulation is then accomplished through changes in posture (e.g., lying vs. standing), piloerection of hairs, the use of thermally advantageous bedding sites (discussed later), and, lastly, by shivering heat production.[21] Except for newly born fawns, deer lack brown adipose tissue for nonshivering thermogenesis.[22]

Deer display yet other behaviors by which they may partially compensate for a negative energy balance during winter. The effects of increased heat loss are mitigated to a large extent by selection within winter concentration areas of both day and night bedding sites that optimize thermal exchange with the environment. Night beds are typically situated under a dense conifer canopy, placed close to the trunk under low branches, and usually surrounded by several other coniferous trees. This combination minimizes convective heat transfer by wind and maximizes back-radiation of long-wave energy from the animal's surroundings. So critical is night-time cover that good bedding sites, once found, are often used year after year. During the daytime, however, deer seek out more open bedding areas adjacent to packed trails, with preference for southerly or westerly exposures, but otherwise without fidelity to particular sites.[23] Here solar gain is maximized as deer bed in shallow depressions in the snow. Skin temperatures of deer so exposed to direct and reflected solar radiation may be elevated 6° or more over that of shaded animals.[24] As the solar angle increases with the approach of spring, this represents an increasingly important contribution to the energy budget of an animal.

FUR—THE MAMMALIAN ADVANTAGE

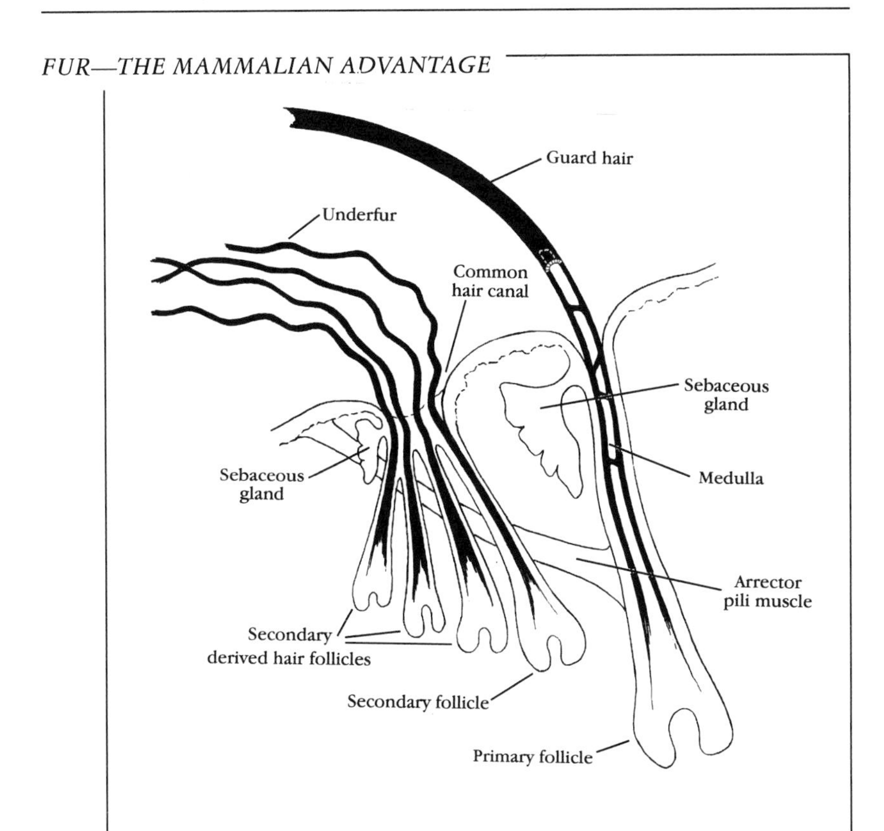

For mammals active above the snow, adaptation to cold often comes down to a single, overriding factor—superior insulation of the winter pelage. The extraordinary cold tolerance of some animals—the marked wintertime depression of lower critical temperatures to −30° C in the moose, −40° C in the arctic fox, possibly even −50° C or below in the caribou—is due largely to seasonal changes in the quality of their fur.

The fur of a mammal consists of two distinct types of hairs: guard hairs are relatively coarse (i.e., larger diameter), long, serve to protect the animal's coat, and usually impart the dominant color to the animal. Underhair is considerably shorter, finer (smaller diameter), and serves primarily as an insulator.

When an animal molts in the fall, the hairs of the new winter coat differ not only in length, diameter, and sometimes color, but also in number. "Derived" hair follicles, associated with the secondary follicles producing underhairs, regress or disappear entirely during summer but become active again in the autumn, multiplying significantly the numbers of hairs produced from a shared canal. Thus, the density of underfur is usually far greater in the winter pelage. Often these underhairs are crimped or wavy, and therefore impart greater thickness to the fur, improving air trapping capability. Guard hairs, too, are sometimes finer in the winter pelage, longer, and often contain large air spaces in the medulla that increases their insulative function considerably.

In addition to hair morphology, several other factors contribute to the insulative quality of the animal's pelage in winter. Keeping the hair dry, an important factor in minimizing thermal conductivity, is largely the function of the sebaceous glands situated adjacent to the hair follicle. These glands secrete oils that are hydrophobic and, thus, help "waterproof" the fur. Erection of hairs, a function of the arrector pili muscles, allows for the entrapment of more air, thus decreasing thermal conductivity further. And finally, a reduction of blood circulation to the dermis of the skin, through constriction of surface vessels, increases tissue insulation immediately below the fur. These characteristics, acting in concert, are truly the mammalian advantage in cold climates.

(*Source:* G. A. Worthy, J. Rose, and F. Stormshak, "Anatomy and Physiology of Fur Growth: The Pelage Priming Process," in *Wild Furbearer Management and Conservation in North America,* ed. M. Novak, J. A. Baker, M. E. Obbard, and B. Malloch (Ontario: Ministry of Natural Resources, 1987). Illustration is a simplified composite drawing from figures 1–3, pp. 827–28, in Worthy et al.)

The telling advantage of these behaviors is seen in late winter when fat stores are depleted, much of the available browse has been consumed, and mobility is hampered by deep snowcover. If fat reserves become exhausted before the animal can regain a positive energy balance, body protein will be assimilated for maintenance energy and, if relief is not soon forthcoming, death will quickly follow.[25] But the effects of a hard winter can be far more subtle for the survivors. Malnutrition is common at the end of the season, and warm-weather recovery of nutritional health may not always be complete. David Mech of the U.S. Fish and Wildlife Service and his colleagues have provided evidence indicating that several hard winters in succession may have cumulative effects on maternal nutrition that strongly influence fecundity and fawn survival. Thus, while predation is often the ultimate cause of deer mortality, winter weather may prove to be the primary regulator of long-term deer population fluctuations in northern regions.[26]

*Elk (*Cervus elaphus*)*

While elk face many of the same problems as deer in snow country, the evolution of gregarious behavior in the former carries implications for overwintering success that go beyond many of the considerations that we have just discussed. Two aspects of this behavior are of particular relevance here:

The herding behavior of elk cows is primarily a predator avoidance mechanism in which the combined senses of many animals are employed in predator vigilance, and each individual in the herd in effect uses the others as cover, in place of vegetation. Thus, herding behavior and the use of open country are complementary, explaining the cow's habitat preferences. The trade-off for increased security, however, is reduced food quality and accessibility in winter. Elk cows rely more heavily on dried grasses at this time of year (fig. 65) and must often dig or "crater" in the snow for them.[27]

The second implication of herding behavior relates to the reproductive biology of the species, and, in particular, to the effect of au-

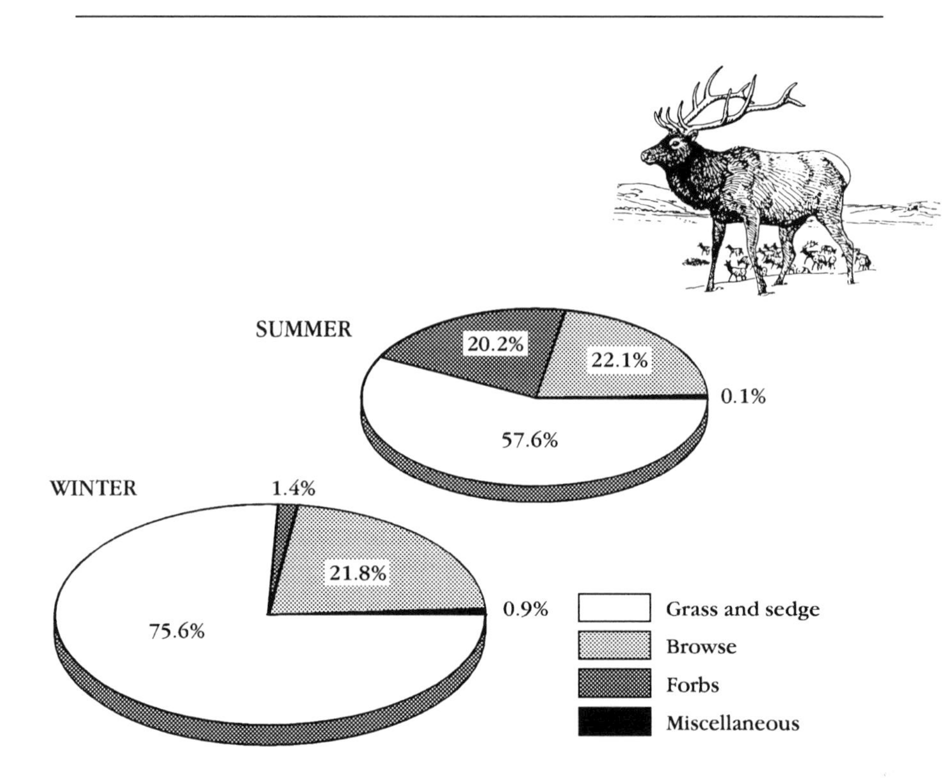

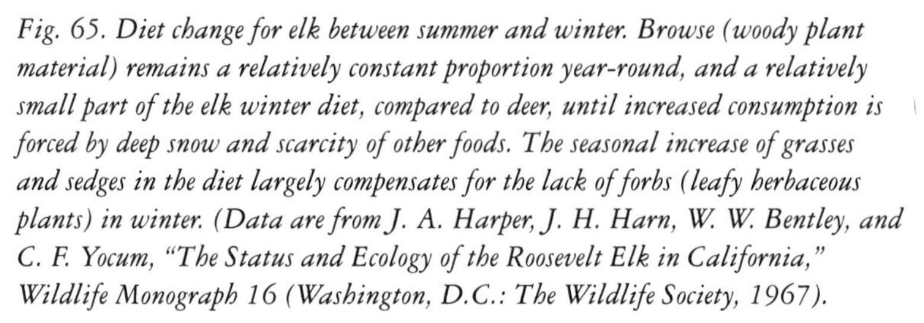
Fig. 65. Diet change for elk between summer and winter. Browse (woody plant material) remains a relatively constant proportion year-round, and a relatively small part of the elk winter diet, compared to deer, until increased consumption is forced by deep snow and scarcity of other foods. The seasonal increase of grasses and sedges in the diet largely compensates for the lack of forbs (leafy herbaceous plants) in winter. (Data are from J. A. Harper, J. H. Harn, W. W. Bentley, and C. F. Yocum, "The Status and Ecology of the Roosevelt Elk in California," Wildlife Monograph 16 (Washington, D.C.: The Wildlife Society, 1967).

tumn rut on the winter survival of bulls. The gregarious nature of the females gives rise to a polygynous mating system in which dominant bulls tend harems of cows against potential competitors. This is an energetically demanding form of polygyny in which the bull must advertise its fitness in order to attract cows, and must then defend its position against challengers, while at the same time keeping the cows from wandering. With little time for feeding, the bull subsidizes this activity with fat reserves accumulated prior to the fall rut—the same fat reserves that must also subsidize dietary deficiencies in winter, as was the case with white-tailed and mule deer. "Caloric bankruptcy" following the rut then often forces bulls to wander, singly or in small groups, in search of high-quality forage such as is often associated with burned areas, sacrificing security for food quality.[28] Thus, elk show strong gender differences in winter feeding and predator avoidance behavior, with the result that bulls tend to suffer greater winter mortality from nutritionally related causes, including increased vulnerability to predators, than do females.[29]

Locomotion in snow is energetically costly for elk as well as for deer (figs. 66*a* and *b*), and accumulating depth usually causes the herd to move to more favorable wintering sites. In some areas elk regularly migrate distances of 80 km or more, from higher summer ranges to valley bottoms. Elsewhere, they may only shift habitat use within their general year-round territory.[30] Early in the winter, when snow is light and sinking occurs to the depth of the snowpack, the greater size of elk (table 8) may confer some benefits in terms of mobility. But locomotion becomes considerably more costly as soon as snow depth exceeds knee height, and once sinking depth exceeds chest height, both deer and elk, if forced to run, resort to an exaggerated bounding gate that increases energy cost exponentially.[31]

Elk have a significant advantage, in terms of knee and chest height, over mule deer of the same age (table 8). In time, however, as snowpack density or hardness increases, sinking depth becomes a function of foot surface area and loading (the weight of an animal divided by the surface area of its four feet), and elk then become somewhat disadvantaged compared to mule deer. Though elk have a total foot surface area nearly double that of the mule deer, their

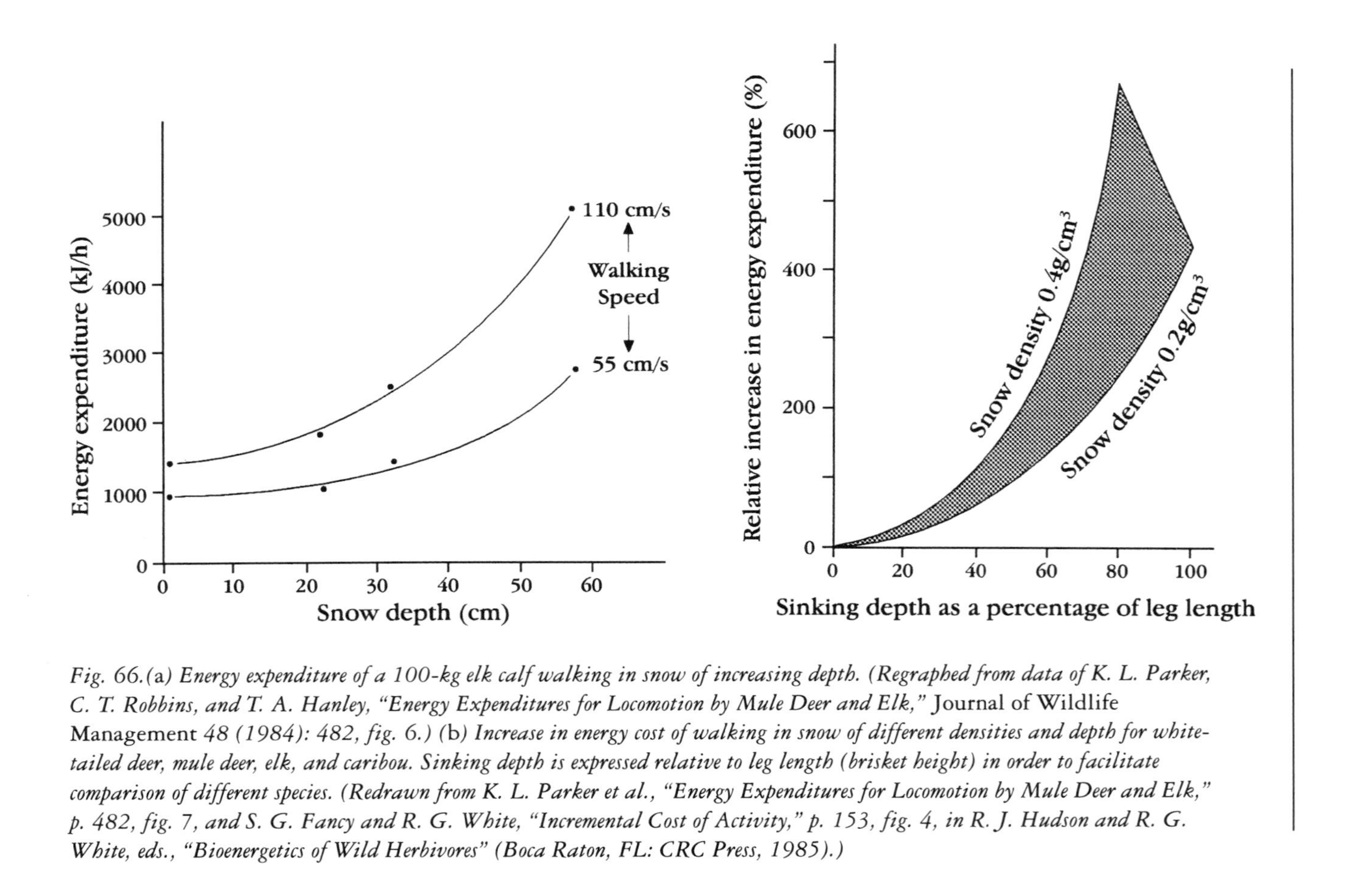

Fig. 66.(a) *Energy expenditure of a 100-kg elk calf walking in snow of increasing depth. (Regraphed from data of K. L. Parker, C. T. Robbins, and T. A. Hanley, "Energy Expenditures for Locomotion by Mule Deer and Elk,"* Journal of Wildlife Management *48 (1984): 482, fig. 6.)* (b) *Increase in energy cost of walking in snow of different densities and depth for white-tailed deer, mule deer, elk, and caribou. Sinking depth is expressed relative to leg length (brisket height) in order to facilitate comparison of different species. (Redrawn from K. L. Parker et al., "Energy Expenditures for Locomotion by Mule Deer and Elk," p. 482, fig. 7, and S. G. Fancy and R. G. White, "Incremental Cost of Activity," p. 153, fig. 4, in R. J. Hudson and R. G. White, eds., "Bioenergetics of Wild Herbivores" (Boca Raton, FL: CRC Press, 1985).)*

Table 8. Critical dimensions of northern cervids for mobility in deep snow

	Mule deer[a]		*Elk*[a]		*Moose*	*Caribou*
	40 kg	*80 kg*	*100 kg*	*200 kg*	*adult*[b]	*adult*[b]
Brisket (chest) height (cm)	53	60	72	77	105	73
Carpus (knee) height (cm)	30	34	42	47		
Foot surface area (cm^2)	36	52	72	102		
Foot loading (g/cm^2)	280	390	350	490	650[c]	190[d]

[a]K. L. Parker, C. T. Robbins, and T. A. Hanley, "Energy Expenditures for Locomotion by Mule Deer and Elk," *Journal of Wildlife Management* 48 (1984): 474–88.
[b]E. S. Telfer and J. P. Kelsall, "Adaptation of some Large North American Mammals for Survival in Snow," *Ecology* 65 (1984): 1828–34.
[c]Value may slightly underestimate "average" loading for adult moose.
[d]Value may slightly overestimate "average" loading for adult caribou.

heavier weight at maturity results in considerably greater loading. Only the elk calf benefits from having the larger feet (table 8). Mule deer are also very adept at collapsing the phalanx and spreading the toes to increase surface area and reduce sinking depth, an ability that elk apparently lack.[32] In addition, elk are less adept at making and utilizing trails in deep snow.[33] While snow depths up to 40 cm generally do not impair elk, this appears to be the limit at which they will crater or paw through the snow for grasses. At greater depths foraging is confined to woody shrubs and twigs of deciduous trees (fig. 67).[34] Thus, in spite of their size difference, elk fare little better than deer in deep snow.

*Moose (*Alces alces*)*

Moose range as far north as suitable habitat allows, showing remarkable tolerance for some of the coldest environments of North America and Eurasia. So well adapted to northern latitudes are they that a winter-acclimatized moose begins to increase respiration and show discomfort when temperatures rise to −5° C. Indeed, it might well be their rather low heat tolerance that sets the southern

Fig. 67. Normally dependent on grasses for a large share of their winter diet, elk are forced to give up "cratering" and forage more on woody plants once snow depth exceeds knee height. (Photo by Bruce Gill.)

limits of their distribution.[35] Moose are also tolerant of relatively deep snow, having the greatest chest height of all the cervids (table 8), and limited data suggest that longer leg lengths are genetically selected through enhanced winter survival (hence greater reproductive success) of individuals in regions of greater average snow depth.[36] It would be hard to argue, thus, that winter has not been the principal architect in the evolution of this somewhat odd looking, but undeniably successful, northern ungulate.

The extraordinary cold tolerance of moose is apparently related to superior insulation. When Per Sholander and his colleagues, in their early arctic investigations, compared the quality of pelts for many species, they found that over the range of insulative values obtained for moose fur, the highest measured exceeded that of all other animals, due in part to the extremely long guard hairs (9 to 10 cm) of the winter coat of moose.[37] Only the fur of caribou, Dall sheep, and arctic fox equaled the quality of moose pelage. With piloerection delayed until air temperatures drop to approximately −25° C, and no increase in metabolic rate with temperatures down to −30° C,[38] the lower critical temperature of moose is estimated at −35° or lower.

Though moose are much better able to cope with snow than deer and elk, they are nonetheless inclined to restrict activities in winter to relatively small areas, conserving energy through limited movement and often utilizing good foraging habitat intensively before wandering on.[39] While individuals may differ widely in their habits, winter home ranges of moose (the area within which an animal generally confines all of its activities) typically contract from around 600 to 1000 hectares in summer, to less than half that in winter. The severity of winter weather may be the most important determinant of actual area usage, with reports of home range size in the literature varying from 225 hectares during mild winters to as small as 2.5 hectares during severe winters or periods of heavy snowfall. At the opposite extreme, however, radio-collared moose in interior Alaska have been found to roam over an area of 300 km^2 during winters of low snowfall.[40] Records of migratory response to snow are likewise varied, but generally indicate only relatively short-distance movements to more favorable food supplies. For some moose, seasonal territory shifts are apparently traditional, persisting through generations as migratory routes are learned by calves accompanying their mothers. Fidelity to these seasonal home ranges is often strong enough that adjacent areas may not be filled when they become vacant.[41] An exception to the limited migratory movements of moose in North America is found in the Seward Peninsula, Alaska, where individuals regularly move a distance of up to 80 km from their summer ranges in the tundra to wintering areas in the lower drainages of major rivers.[42]

Favorable winter habitat for moose is determined primarily by concentration of food supplies, with protective cover apparently of secondary importance. Throughout the range of moose, riparian or

Fig. 68. With their superior insulation, moose are less dependent on conifer stands for thermal protection in winter. They will, however, switch to conifer browse when snow becomes too deep in open riparian areas, in order to save energy expended on foraging. (Photo by Bruce Gill.)

streamside habitats are strongly preferred during winter for their high productivity of shrubs, particularly willow. Where access to riparian areas is limited, moose exploit the early successional stands of paper birch and quaking aspen found in forest openings. In eastern Canada and northern New England, for example, balsam fir stands mixed with birch and aspen, where logging or fires have created gaps in the overhead canopy, are essential winter habitat for moose. Coniferous forest cover is much less important in this case for its thermal protection, but in times of deep snow moose will shift from deciduous to coniferous browse to minimize foraging effort (fig. 68).[43]

The importance of high-density browse availability has less to do with the large food requirements of moose than it has to do with the opposite, expense side of the energy budget. Like deer and elk, moose are limited in their energy intake by the amount of time required for the digestion of poor-quality food, and are not likely to be able to satisfy all their energy needs from browsing, even under the best of circumstances. But a high density of suitable shrubs means that a ruminant can ingest a large amount of material in a relatively short time, satisfying its bulk capacity with minimal effort, and then bed down to digest the food, conserving energy that would otherwise be spent on additional foraging. At low browse densities, the animal would either have to move further to find high-quality food, or become much less selective in what it eats.

Since virtually all of the energy expended by a moose in winter is directed at feeding, activity levels of moose show a strong inverse relationship with food density. In Denali National Park, Alaska, where browse is abundant, moose devote slightly more than 6 hours a day to foraging and spend nearly twice that amount of time ruminating. Foraging bouts usually last less than an hour, so that moose there generally exhibit six cycles of feeding and bedding each day during winter.[44] A detailed activity budget for a free-ranging moose confined within a 65-hectare enclosure at the Ministik Wildlife Research Area near Edmonton, Alberta, is shown in figure 69. The somewhat greater proportion of time devoted during February to feeding (compared to Denali moose) might in this case reflect either a lower density of browse within the 65-hectare enclosure, requiring more effort to fill the rumen, or decreased foraging efficiency (e.g., greater selectivity on the part of the animals) due to the fact that the captive moose were intermittently fed pelletized food during nonexperimental periods, to help maintain them. It is unlikely that differences in ambient temperature during the respective studies influenced the amount of time spent feeding—moose are so cold tolerant that in Denali their activity patterns were entirely independent of winter weather.[45]

The total quantity of food required by moose in winter is impressive. Relative to their body weight, moose have a slightly lower resting metabolic rate than other cervids, ranging between 430 and

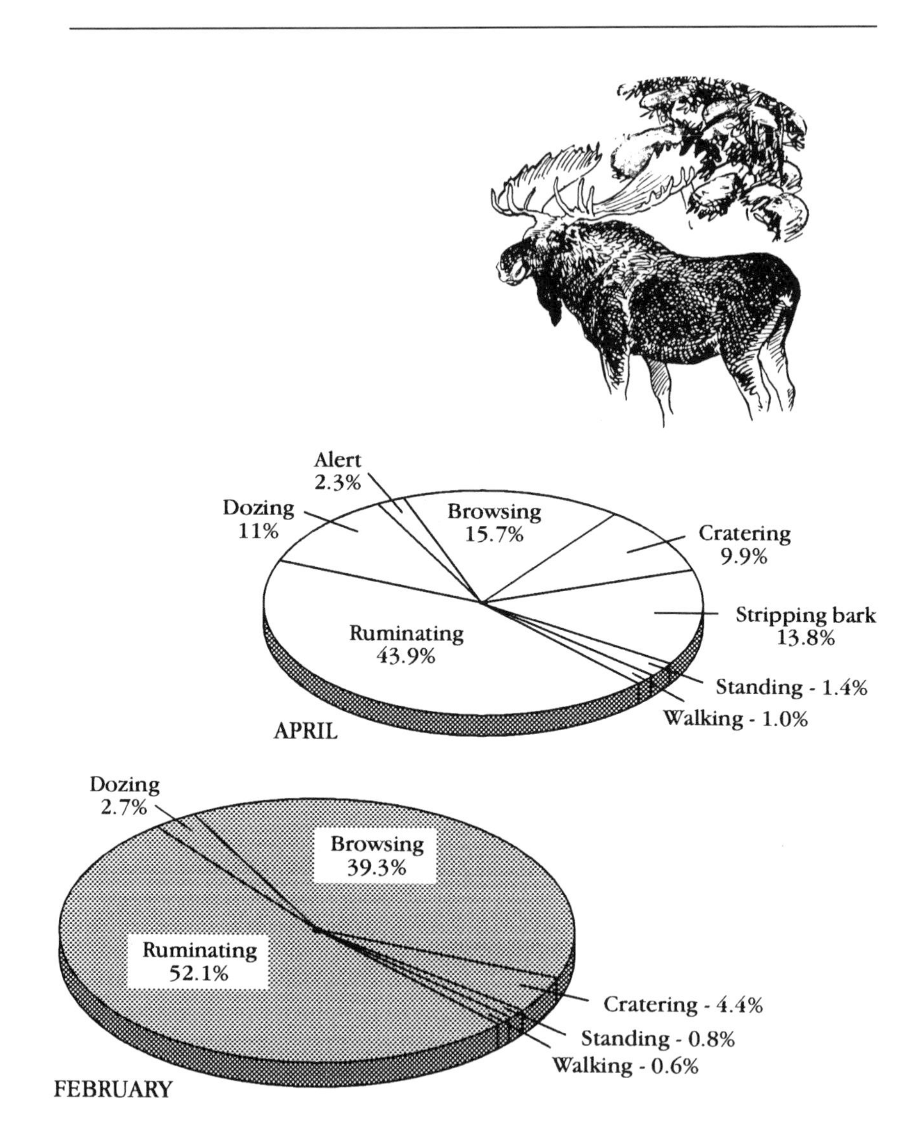

Fig. 69. Daily activity budget (percent of day engaged in various activities) for a cow moose within a 65-hectare enclosure in Alberta, Canada. See text for details. (Data from L. A. Renecker and R. J. Hudson, "Seasonal Activity Budgets of Moose in Aspen-Dominated Boreal Forests," Journal of Wildlife Management *53 (1989): 298.)*

452 kJ per kg^{75} body weight per day[46] (expressing weight to the .75 power reflects the "metabolic" size of the animal, as discussed at the beginning of this chapter). Due to their large size, however, this translates into a substantial need. With an additional energy "tax" for foraging, conservatively estimated at 55% of the resting metabolic rate, plus the incremental cost of walking in snow with, say, a sinking depth of 60% of chest height (see fig. 66), the daily energy requirement of a 400-kg moose might well exceed 62,000 kJ, requiring roughly 15 kg of fresh browse every day. This level of consumption is probably seldom sustained, however, as evidence suggests that moose may voluntarily reduce energy intake in winter, even when offered high-quality food.[47] In a recent study of free-ranging moose in Norway, cows early in the winter gained a surplus of energy from browsing in better quality habitats, but by late winter intake was down considerably, resulting in an overall energy deficit of 25 to 30% for the season.[48] Thus, moose also depend on fat reserves accumulated earlier to subsidize winter food requirements.

*Caribou (*Rangifer tarandus*)*

The many recognized subspecies of caribou (called reindeer in Eurasia and also when referring to domestic caribou) occupy a variety of habitats from boreal forest to high arctic tundra. Some are noted for long-distance migrations between arctic coastal calving grounds and wintering habitat in the forest-tundra transition zone, while others remain sedentary year-round. Historically, woodland caribou were found as far south in the New World as northern New England, while at the other extreme, the Peary caribou today push their range limits to the northernmost parts of the Canadian high arctic on the Queen Elizabeth Islands. Only one other ungulate shares the distinction of ranging this far north, and that is the musk ox.

The caribou's ability to cope with the climatic rigors experienced at the extremes of its geographical range are truly outstanding. Its exceptional foot surface area, relative to its size, gives it superior mobility in the snow (a subject we will return to later), and the excellent insulative qualities of its pelage, coupled with effective peripheral cooling, minimizes heat loss to its environment. The long guard hairs of the caribou fur are thick and hollow, with air spaces separated by thin septa,[49] greatly increasing their effectiveness in reducing conductive heat loss (see box on "Fur—The Mammalian Advantage"). Constriction of surface blood vessels, especially in the extremities, also increases tissue insulation. This vasoconstriction, combined with effective heat exchange between veins and arteries in the legs, enables caribou to maintain a 30° C temperature gradient between the foot and trunk,[50] further reducing heat loss to their environment. Metabolically, however, caribou differ little from the other cervids, except perhaps in showing greater variability of metabolic rates reflecting the wide geographic range of the species. At one end of the spectrum, the Svalbard reindeer have a resting metabolism measured at 453 $kJ/kg^{.75}/day$, similar to the moose,[51] while at the other end, resting metabolic rate of 617 $kJ/kg^{.75}/day$ has been estimated for caribou in Denali, Alaska.[52]

With the exception of foot morphology, the physical and metabolic characteristics just described differ only to a degree among the northern cervids. It is an important degree in this case, however, driving the lower critical temperature of the caribou significantly downward. In sum, these characteristics confer extraordinary cold tolerance to the caribou, its lower critical temperature possibly well below −40° C. In one study, a 9-month-old caribou calf showed no increase in metabolic rate at temperatures down to −55° C.[53]

Apart from such noteworthy cold tolerance, the principal difference in overwintering strategy between caribou and the other cervids lies in their food habits and the manner in which they satisfy their daily energy requirements in winter. Throughout their range, caribou feed on a wide variety of plant material, including woody twigs, evergreen leaves, grasses, sedges, and fungi. However, their preference for lichens is notable and unique among the cervids, and the key to their survival in many regions.[54] Lichens, both ground-covering and arboreal, are widespread in tundra and boreal forest environments, and though their productivity is rather low, they are a ready source of carbohydrate with much less cellulose to contend with. The gut flora of the caribou is uniquely adapted to this diet.

Caribou are apparently able to locate lichens beneath the snow by smell and dig them out with their broad, sharp-edged front hooves, provided the snow is not excessively deep. They are able to break through thin frozen or wind-packed crusts with no greater energy expenditure than that required for walking on snow-free ground.[55] In boreal forest regions, however, where trees support much lichen growth, caribou will shift to arboreal lichens readily if the snow crust exceeds a hardness of 50 g/cm^2 or the snowpack depth exceeds 60 to 75 cm.[56] In the arctic, caribou have been reported to break through crusts approaching a hardness of 10,000 g/cm^2,[57] but will spend increasing amounts of time in search of soft snow as the energetic cost of cratering becomes too great.

Caribou are well equipped for moving efficiently over snow-covered ground, having a considerably lower foot loading than any other northern cervid (see table 8). The hoof of the adult caribou is as wide or wider than it is long, providing a nearly circular bearing surface of 12 to 14 cm diameter. In addition, the dewclaws of the caribou are large and set back a greater distance from the hooves than is typical of most ungulates. These provide additional bearing, particularly for the front hooves, as the foot is planted at an exaggerated angle from the vertical.[58] The foot loading given in table 8 is about 25% higher than values for adult caribou reported elsewhere,[59] and may reflect a difference in measurement of the actual bearing surface in snow. It is noteworthy, too, that the feet of caribou stay flexible at cold temperatures because the fatty tissues deposited therein remain soft, even though marrow fat in the leg is solid.[60]

These advantages notwithstanding, the greatest measure of energy conservation in winter still comes with voluntary reduction of activity, a seemingly universal behavioral strategy among the northern cervids. Svalbard reindeer, arguably the best suited of all caribou to the rigors of the high arctic, show a characteristic lethargy in winter, spending 45% of each day lying, sometimes for uninterrupted periods of 5 to 6 hours.[61] The energetic advantage of this became clear when L. C. Cuyler and N. Øritsland of Norway measured the lying and standing metabolic rate for subadult reindeer and then applied estimated energy costs to the activity budget of free-ranging animals (fig. 70*a*). Their calculations suggested that the long resting times saved 25 to 35 days worth of forage over the course of the winter, as compared to the needs of caribou that were more active. But Svalbard reindeer also have a remarkable ability to store fat, which is an essential aspect of a voluntary reduction of activity. At the start of winter, fat content in these animals is often 30 to 40% of the lean body weight, a considerably greater reserve than is found in most caribou, though by the end of the season their total weight loss may approach 50%.[62] Even the Denali caribou, with their somewhat higher daily energy consumption, show a remarkable degree of conservation in winter. Combining the benefits of superior insulation, a lowered resting metabolic rate, and supplemental fat reserves, these animals expend in winter only slightly more than half the energy required during the summer insect season (fig. 70*b*).[63]

In spite of these advantages, caribou must still forage, and finding food can be energetically expensive. Cratering in hard or deep snow becomes particularly costly, and since the animal usually obtains only a small amount of lichen with each effort, several dozen craters may have to be dug to satisfy their daily needs. Where lichens are sparse, more craters must be excavated, whence more food is needed to offset the added effort. Lichens are slow growing, requiring decades to establish full cover, but caribou are cursory harvesters, always moving and, hence, never cropping the lichens fully in one place. Time and space, thus, are important dimensions of the caribou's habitat, and fire and human activity (the two often related) have had great impact on both in the last century, with the result that herd sizes are diminishing in many parts of the North.[64] In spite

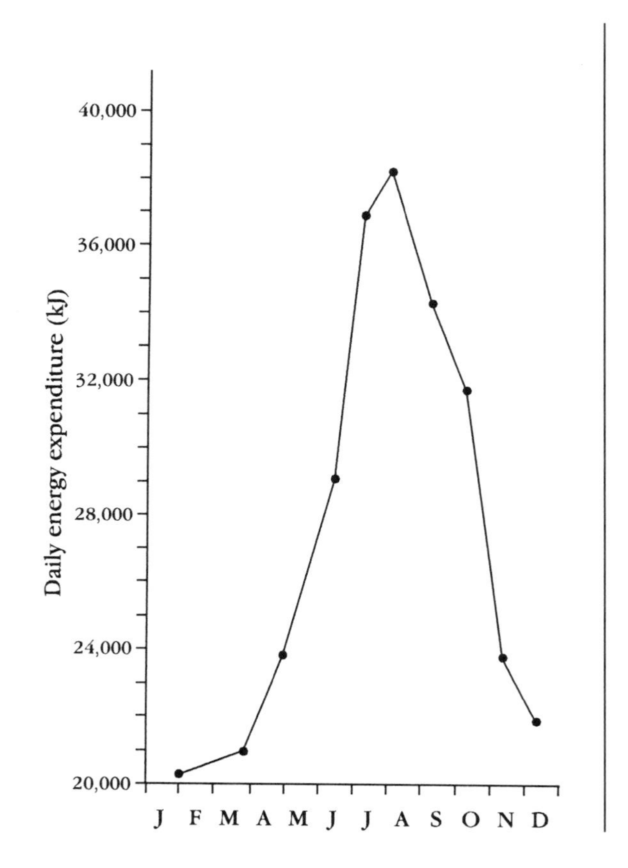

Fig. 70.(a) *Daily activity budget and energy expenditure for a subadult Svalbard reindeer in winter. See text for details. (Data are from L. C. Cuyler and N. A. Øritsland, "Metabolic Strategies for Winter Survival by Svalbard Reindeer,"* Canadian Journal of Zoology *71 (1993): 1787–92.)* (b) *Daily energy expenditure throughout the year for Denali (Alaska) caribou. Lowest energy expenditure occurs in mid to late winter. The rising limb of the annual curve coincides with the beginning of spring migration (April-May), the calving and postcalving season (June), and summer migration (early July). Highest daily energy expenditure occurs during the peak of the insect season, from mid July to the end of August. The falling limb of the curve then coincides with the beginnings of fall migration (early September), the rutting season (October), and the end of migration in early winter. (Regraphed from R. D. Boertje, "An Energy Model for Adult Female Caribou of the Denali Herd, Alaska,"* Journal of Range Management *38 (1985): 468–73.)*

of their remarkable adaptations to life in the cold, caribou still suffer high mortality when winters are hard and the lichen cover thin.

SEMIAQUATIC MAMMALS

The semiaquatic mammals of cold regions—muskrat, beaver, river otter, mink—face the particularly difficult problem in winter of adjusting continually to both aquatic and terrestrial environments. We are familiar now with the challenges of terrestrial habitats in winter. We have also considered many aspects of the aquatic environment under ice (chapter 5) and are reminded that air-breathing, homeothermic mammals are still somewhat alien to this environment and face a different set of problems in winter than the aquatic organisms that we discussed earlier. With the thermal conductivity of water many times greater than that of air, a significant increase in the potential for heat loss occurs whenever an animal enters the water. And while the semiaquatic mammals show a seasonal increase in fur thickness comparable to other land mammals, the displacement of air entrapped in the pelt with underwater compression of the fur dramatically reduces the effectiveness of their insulative coat. Thus denied some of their best defense against low temperatures, mammals foraging under the ice must rely upon additional behavioral and physiological adjustments to prevent immersion hypothermia, while still maintaining their terrestrial adaptation. Each of these species has evolved a slightly different approach to the problem.

*Muskrat (*Ondatra zibethicus*)*

Muskrats generally exploit shallow waters with an abundance of emergent vegetation, which they utilize both for food and for construction of shelters. Unlike beavers, muskrats do not maintain food caches in winter (except in unusual circumstances where they may stockpile windfall food resources such as ear corn[65]), but rather, they forage daily throughout the winter, feeding primarily on the rootstocks of plants like cattails and bulrushes, and on the leafy parts of submerged aquatic plants.

Prior to freeze-up, muskrats construct and maintain feeding shelters at varying distances from their lodges to facilitate winter foraging. Feeding shelters resemble lodges (living shelters), but are generally less than half the dimensions[66] and are constructed of emergent plant materials with a feeding platform situated above water. A plunge hole through the ice is kept open by the muskrats. Another feeding structure, known as a "push-up," consists of accumulations of submergent vegetation, either wedged up through cracks in the ice or piled up in a somewhat fragile, frozen dome above a plunge hole. As will be seen, these unique winter structures not only enhance foraging efficiency but help solve two important problems for the muskrat: one is related to heat balance and the other to aerobic diving endurance under the ice.

When a muskrat leaves the warmth of its lodge on a feeding foray, it is faced, as already noted, with a dramatic increase in the rate of heat loss by conduction from its body core—much more so in near freezing water than it would experience if it were foraging on land in air of considerably lower temperature. The extent of the problem can be appreciated by substituting into the equation on p. 95, approximate values reflecting the change in temperature gradient (T_b-T_a), a reduction in thickness of the insulative layer (d), and a concomitant increase in thermal conductivity (k) of the wetted fur (about 4 times that of dry fur in the case of the muskrat[67]). For the animal, the result holds two important repercussions. The first is difficulty in maintaining body temperature. When Robert MacArthur at the University of Manitoba actually measured the abdominal temperatures of muskrats diving in water at 3° C, he found that they underwent systematic cooling at a rate of 0.14° C per minute. It appeared that the animals could not (or did not) match the high rate of heat loss to the cold water with an equivalent amount of heat production. The second, related problem is that in water the lower critical temperature of the muskrat is effectively raised more than 20° to just over 30° C,[68] potentially increasing the energetic demands of the animal by a significant amount.

Taken together, these two problems should prove forbidding to the muskrat. Given the cooling rate that MacArthur observed, it seems that muskrats diving under the ice would likely become severely hypothermic within a short period of time. However, in previous studies of free-ranging muskrats in winter, MacArthur, using radio telemetry, discovered that the animals rarely experienced an abdominal temperature decline of more than 2°, even during foraging excursions under the ice in which the animals were away from the lodge more than 40 minutes[69] (a cooling rate of 0.14°/min would predict a body temperature drop of 5.6° in 40 minutes). Clearly there is more to the situation than revealed by simple energy-budget analysis. Subsequent investigations, combined with earlier field observations, have revealed a number of behaviors by which this apparent discrepancy can be resolved.

Muskrats are meticulous groomers, devoting considerable time within the lodge, either singly or communally, to spreading oils from the sebaceous glands throughout their entire pelt (see box on "Fur—The Mammalian Advantage"). An additional source of lipids for treatment of the pelage is provided by the Hardarian gland, situated in the orbital socket of the eyes and discharging into the corner of the eye as well as through the nasal duct. The hydrophobic secretions from these glands retard wetting of the fur, thus aiding in the retention of more air within the pelage and reducing water contact with the skin when the muskrat dives. This entrapped air may amount to 20% of the dry volume of the animal.[70] When muskrats return to the lodge with depressed body temperatures, they again engage in vigorous grooming, perhaps as much to generate heat and stimulate peripheral circulation, as to reapply oils to their fur (interestingly, animals that are prevented from grooming are slower to rewarm[71]). The importance of this grooming behavior in retarding immersion hypothermia is indicated by experimental observations of increased heat loss in cold water from animals that had been shampooed to remove the lipids, and from animals that have had the Hardarian gland removed surgically.[72]

In addition to judicious preparation of their fur, MacArthur has also shown that prior to entering the water, muskrats deliberately raise their core temperature by as much as 1.2°, a behavior that he observed in wintertime only. The mechanism by which this predive hyperthermia is accomplished is unknown at present, but curiously (though perhaps unrelated) MacArthur notes that he frequently observed muskrats to hyperventilate just prior to diving, nearly doubling their normal breathing rate while sitting motionless at the edge of the plunge hole.[73] In the water, this added thermal inertia would give the animal a theoretical advantage of 9 minutes under the ice, based on the rate of heat loss cited earlier, before they would cool to normal body temperature.

The combination of predive hyperthermia and relatively effective retention of air in the pelage, coupled with highly efficient vasoconstriction and countercurrent heat exchange in peripheral tissues such as the tail[74] (refer to pp. 115–17 for discussion of this mechanism), is apparently sufficient to retard development of anything more than mild hypothermia on short foraging trips under the ice. Assuming that muskrats can readily withstand a 2° drop from normal abdominal temperature (since MacArthur frequently observed this in winter), it appears that they could safely endure 25 minutes of submersion, at the cooling rate observed by MacArthur, without increasing heat production, even though they are operating considerably below their lower critical temperature. Though muskrats are well endowed with interscapular brown fat, they apparently defer major nonshivering thermogenesis (pp. 106–7) until they are out of the water, and only on longer excursions do they rely on feeding shelters or push-ups for brief periods of rewarming.[75]

If cold-water immersion would seem a constraint to the foraging activity of aquatic mammals in ice-covered ponds, so, too, would limited access to air. Based on measurements of total oxygen storage capacity and estimates of oxygen consumption rates while diving, the underwater endurance of muskrats should, at best, be no more than 58 seconds[76]—far less than the animal's immersion tolerance (it is presumed to be advantageous to the animal to stay within the limits of aerobic activity in order to minimize disturbances to blood chemistry—e.g., lactic acidemia—and reduce recovery time). At an average underwater swimming speed of .76 m/s, 58 seconds translates into a straight-line distance of 44 meters,[77] a rather limited range for muskrats occupying a communal lodge and competing for food. Here, the advantage of a system of above-ice feeding shelters seems particularly obvious, and often the average distance between such shelters in northern marshes is within this calculated diving range.

The aerobic dive limit of 58 seconds actually represents a 41% increase in underwater endurance over that observed in summer, extending the muskrat's diving range under ice by about 13 m. This rather dramatic improvement in winter performance under ice is related to a substantial increase in the animal's ability to store oxygen in winter. Several known factors contribute to this, including a seasonal increase in blood volume (the major storage compartment for oxygen), enhanced affinity of hemoglobin for oxygen in winter, and an increase in muscle myoglobin.[78] Even with this advantage, however, feeding shelters are located often enough at distances exceeding the calculated aerobic range of muskrats to suggest still other mechanisms for prolonging foraging time under the ice.

Underwater routes regularly traveled by muskrats in winter are frequently delineated by bubble trails of gasses released by the animals and subsequently trapped under the ice (fig. 71). In the early literature, investigators often noted, whenever clear ice permitted observation, that swimming muskrats would occasionally pause at some of these bubbles, where they appeared to inhale air from them. The implication, of course, was that the bubbles provided a usable source of oxygen under the ice.

It was not until 1992, however, that Robert MacArthur experimentally confirmed what others had suspected. He constructed a diving chamber that was covered with Plexiglas to entrap air bubbles released by a swimming muskrat, just as a natural ice cover would, and fitted the chamber with a movable guard by which he

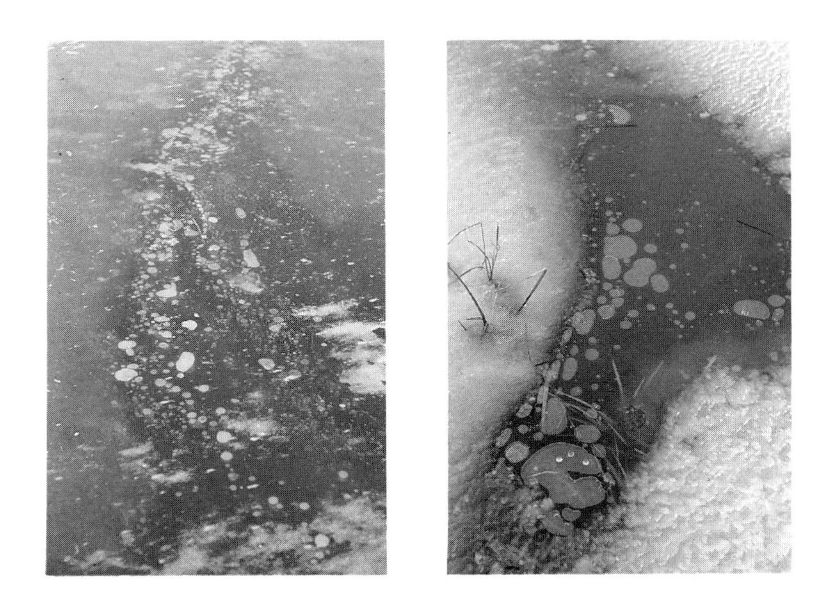

Fig. 71. Under-ice bubble trails created by muskrats as they forage from a bank burrow. (Photos by Robert A. MacArthur.)

could either allow or deny access to the air bubbles as they accumulated. By means of a gas-sampling syringe mounted on the tank cover, he could also measure the volume of air released by the muskrat and analyze its content for oxygen and carbon dioxide. A similar, larger tank was provisioned with an underwater feed bed of aquatic plants to simulate more natural foraging conditions. With this apparatus, MacArthur was able to observe muskrats closely and time their underwater activities with and without access to accumulated air bubbles.

Under the reasonably natural conditions imposed by this experimental setup, muskrats on each dive released about 15% of the estimated air stored in their lungs and pelage, with the released bubbles containing between 11.25 and 18.51% oxygen. When muskrats were later allowed access to the accumulated bubbles, they frequently interrupted their swimming, penetrated the air pockets with their nostrils, and inhaled, visibly reducing the size of the bubble. Access to air pockets did not increase the frequency of underwater forays, but use of this source of oxygen nearly doubled voluntary submergence time, compared to animals that were denied access to the accumulated bubbles, and substantially increased the number of excursions that exceeded in duration the calculated aerobic dive limit of the muskrats.[79] Why muskrats gave up oxygen in the first place (the bulk of the air released was from the lungs) remains unanswered. The amount exhaled was not sufficient to appreciably affect their buoyancy and, hence, energetic requirements for diving. One possibility is that with the high solubility of carbon dioxide in water, expired CO_2 would diffuse rapidly into the surrounding water, and thus, releasing air from the lungs could serve to reduce detrimental CO_2 buildup—a convenient mechanism of extending dives if the released oxygen is available later. MacArthur suggests that the habit of continually releasing small amounts of air from the lungs might also facilitate slowing of the heartrate during cold-water submergence.[80]

The ability to scavenge oxygen from gasses previously released by the animals themselves, or to recover oxygen from bubbles released over beds of photosynthesizing plants (see discussion on p. 161), would be of obvious advantage to an aquatic mammal under the ice, especially when coupled with increased oxygen storage capacity in winter. The added ability to compensate for increased heat loss to the water by means of predive hyperthermia, and the advantageous use of shelters above the ice for warming and feeding, demonstrates a complex and remarkable adaptability on the part of the muskrat to one of the more challenging winter environments.

The one exceptional difficulty that muskrats cannot deal with effectively is getting frozen out of their shallow marshes during extended hypercold, when the water may freeze to the bottom. Errington, in his extensive work with muskrats in the 1940s and 1950s, records such occurrences in Iowa and South Dakota, when even fishes that had sought their last refuge in muskrat channels in the bottom mud became encased in ice and were gnawed out by the hungry muskrats.[81] Food shortage does not seem to be the crucial problem in such times, however, as muskrats are capable of surviving on vegetative matter protruding above the snow if forced out. Rather, predation, principally by mink accessing poorly constructed lodges from above the ice, and a lack of effective shelter present the greatest problems. Errington has noted that muskrats may remain alive for weeks in midwinter, in drought-exposed marshes, as long as they have access to better lodges.[82] Recent studies of muskrat lodging habits in Saskatchewan have shown that burrows excavated into shoreline banks had a higher probability of remaining occupied during winter, probably for these reasons.[83]

*Beaver (*Castor canadensis*)*

Though beaver occupy much the same habitat as muskrat, they have evolved a markedly different behavioral pattern in dealing with ice cover, particularly with regard to their winter feeding habits. Having become primarily specialist feeders on the bark of trees and shrubs, northern beavers are usually denied access to terrestrial food supplies by persistent ice cover, and so must rely instead on limited underwater caches and occasional submergent aquatic plants.

Winter food caches consist of little more than a pile of stems and branches, mostly of broadleaf trees and shrubs but occasionally including conifers, anchored underwater close to the lodge entrance (fig. 72). Stems may number from 200 to several hundred, depending on the size of the colony, and are mostly small (under 6 cm diameter), though infrequently a manageable-sized tree is included.[84] The cache is constructed in autumn and is usually secured by "planting" the cut stems into the bottom mud. Occasionally floating rafts of vegetation are assembled, that eventually become anchored in the ice. Though rarely described in the literature, it has been reported that beaver will place less desirable material such as alder or peeled aspen in the upper portion of the raft where it freezes into the ice and becomes unavailable. More desirable species are dispersed throughout the lower portion of the raft where they remain accessible in winter.[85] How commonly this is done is not

Fig. 72. Winter food cache prepared in autumn by beaver. Leaves are still present on the fresh saplings piled close to the lodge. (Photo by Peter Marchand.)

known. Notable deviations in caching behavior are found among populations of beaver inhabiting rivers that undergo wide seasonal variations in flow. A pronounced wintertime drop in water level often leaves hanging ice, and beaver are able to forage in the air space ("sushinetz" in Russian literature) so created.[86]

Though the literature suggests considerable geographic variation in winter feeding activity, adult northern beaver often show a reduction in energy intake even when food is available. In a unique study during the mid-1960s, N. Novakowski noted that kits and yearling beaver in the Mackenzie River drainage in northern Canada gained weight during the winter, while adults lost weight. When he analyzed the entire food caches of five different colonies for available energy, he found that two of the five were marginally sufficient to meet the calculated needs of the respective colony, while the other three fell far short, even though food available for caching was very abundant throughout his study area[87] (later refinements in estimating daily metabolic requirements for beaver in winter suggest that Novakowski's colonies may have experienced an even greater energy deficit than he predicted[88]). Given that the beaver populations in his study area had been stable for 10 years, Novakowski concluded that the caches were adequate only to support the number of young in the colony and that adults overwinter "in relative poverty," enduring the deficit on fat reserves. This conclusion finds some support in observations that starvation among beaver colonies is rare, and that the tail of the beaver, a fat storage depot, often undergoes a reduction in dimensions during winter.[89] While Novakowski's study suffers from some of the limitations that characterize many field investigations, his detailed analysis of winter food caches still stands as a unique effort.

Shortly after Novakowski published his results, M. Aleksiuk and I. Cowan reported on similar investigations in the Mackenzie Delta area in the Northwest Territories, in which they found winter growth cessation in immature as well as adult beaver. While their observations were not entirely in disagreement with Novakowski's (i.e., that energy intake during winter was minimal), their own interpretation took a somewhat different slant—namely, that regulation of energy *expense* during winter is under the control of some intrinsic physiological mechanism, and that large winter fat stores indicated a level of expenditure at or below energy intake. Winter growth cessation, they observed, was displayed even among captive animals given unlimited access to food (though not so with beavers from California).[90] This led them to postulate a winter metabolic depression in northern beaver, an idea that gained rapid acceptance initially, apparently more for its intuitive appeal than for experimental verification, but that has since become the source of considerable debate.

In recent years, efforts to substantiate a hypothesized winter metabolic depression have centered on seasonal measurements of abdominal temperature in free-ranging beaver, with the principal argument being that changes in metabolic rate, whether associated with sleep, shallow torpor, or hibernation, are nearly always accompanied by changes in body temperature,[91] and, therefore, should be observable by radio telemetry. While several researchers have been successful in instrumenting beavers, their efforts so far have not clarified the issue. Douglas Smith and his colleagues at Michigan Technological University reported a distinct 1° winter depression in body temperature of an adult and yearling beaver, noting that even with an activity level under the ice comparable to that observed in autumn, daily minimum body temperatures were consistently lower in winter (the one instrumented kit in their study showed low activity and maintained a higher—normal—body temperature throughout winter, and kits in this colony were found generally to gain weight over winter).[92] These investigators proposed that body temperature was precisely regulated year-round and that the lower temperature in winter was part of a strategy to reduce energy expenditure when resources were scarce. In southeastern Manitoba, Alvin Dyck and Robert MacArthur failed to find similar evidence of a body temperature drop, but acknowledged that geographic variation in habitat quality and, in particular, winter food availability could be responsible for the observed differences in the two colonies.[93] Thus, while the energetic advantages of a winter metabolic depression are understandable, the generality of such a strategy among northern beaver remains unproven.

Studies of the preceding nature have, however, revealed considerable differences in winter activity levels among beaver from geographically separated colonies. Novakowski had previously reported, from analysis of trapping records, that ice-bound beaver left the lodge only once every two weeks on average. While that may be typical for the northern Mackenzie River area, more recent radio telemetry studies in other regions have indicated considerably greater under-ice activity, ranging from weekly to daily excursions, with forays lasting from less than 5 minutes to more than 40 minutes.[94]

Latitudinal differences in the level of under-ice activity could be related to differences in light conditions, but not in temperature, since water temperature will vary by no more than 4° under ice cover (see pp. 144–47). With regard to light response, however, studies have revealed yet another enigma in the winter behavior of beaver. In virtually all investigations by remote sensing technology, beaver have been found to deviate from normal 24-hour circadian rhythms under the ice and exhibited free-running activity patterns, with a periodicity of from 26 to 29 hours. In some cases, activity cycles were synchronized between individuals of the colony, but often they were not.[95]

Normally, light (specifically changes in intensity at sunrise and sunset) acts as the external environmental cue that serves to entrain daily activity rhythms of an animal (i.e., set the internal "clock") to a 24-hour period. For this reason, the frequent observation of free-running behavior in winter has led often to the conclusion that beaver experience continuous darkness under the ice. Yet we know from earlier considerations of light penetration through snow and ice (pp. 32–36) and confirmation of active photosynthesis in submergent aquatic plants in winter (p. 161) that "continuous darkness" in the aquatic environment is a condition likely to be experienced only at very high latitudes. Why beaver in southern Manitoba, northern Michigan, eastern Massachusetts, and southern Quebec[96] deviate from 24-hour activity rhythms under ice cover, while other organisms remain responsive to low light, is not clear, Perhaps for beaver, as with reindeer and humans, there exists a threshold light intensity, below which the dim light under ice fails to inhibit synthesis of melatonin, the hormone secreted by the pineal gland that serves as the light-measuring "clock setter" in mammals. The ecological benefits of free-running behavior are unclear, too. Allowing circadian rhythms to drift out of phase with external cycles and the activity patterns of predators probably carries little risk under the ice, but what might be the advantage? Perhaps the desynchronization of foraging excursions among members of a colony assures that the lodge will always remain warm (see fig. 33, p. 104).

Given that beaver do forage to one degree or another under the ice, what can be said of their tolerance to cold water immersion? As with the muskrat, much of the beaver's thermoregulatory strategy in winter is based on behavior, rather than on cold endurance. Though endowed with dense pelage consisting of long guard hairs (up to 50 mm) covering a thick underfur of wavy hairs that impart a downy softness to the coat, the beaver still suffers a significant loss of insulation when diving. Its lower critical temperature in water ranges between 20 and 25° C,[97] only slightly lower than that of the muskrat. Without thermoregulatory compensation while in the water, the beaver also experiences a decline in body temperature, but due to its larger size (coupled with efficient countercurrent heat exchange in the tail and hind limbs) gains considerably longer immersion time without suffering excessive hypothermia. The per-gram cooling resistance of the beaver is 50 times that of the muskrat, while its whole-body cooling rate is 3.5 times slower (see "The Size Trade-Off" on p. 119). The beaver also elevates its body temperature slightly (0.22 to 0.64° C) before diving in cold water,[98] which, with its larger size, gives it about the same thermal inertia and time advantage underwater as the muskrat.

The greater passive resistance of the beaver to immersion hypothermia would seem of advantage to it only if its foraging excursions under the ice were of long duration. Though rarely recorded in the literature, the beaver has been known to remain away from the lodge for lengths of time exceeding its aerobic endurance, suggesting that the beaver either utilizes air spaces created by receding water levels or, like the muskrat, is also able to scavenge oxygen from air bubbles under the ice.[99] However, by its food caching behavior, its reduced energy expenditure, and its communal lodging, the beaver behaviorally avoids many of the problems associated with cold-water immersion and oxygen limitation, spending much of the winter within the lodge, in a microclimate where temperatures may seldom fall outside of the animal's thermoneutral zone (fig. 33, p. 104).

*River Otter (*Lutra canadensis*)*

While the river otter is almost entirely dependent in winter upon the food resources of the aquatic environment, it is not nearly as well suited behaviorally to a life under ice as are the muskrat and beaver. An active carnivore, the otter must cover considerably more territory in its hunting, crossing the boundaries of terrestrial and aquatic environments many times daily. This dual requirement of land travel between foraging sites, often in deep snow, and of obtaining food in water, often restricted by the presence of ice, demands the utmost of anatomical and physiological adaptation. Yet, in the final analysis, habitat character may be the sole determinant of the otter's overwintering success. So critical is simultaneous access to water and land shelter for the river otter that good winter habitat alone may establish the carrying capacity of a region for this species.[100]

Although notoriously eclectic in their feeding habits, fish comprise the bulk of the river otter's diet throughout its range (see fig. 73 on p. 229 in next section). This is particularly so during the winter months in northern areas, when other components of the otter's usual diet, such as insects and amphibians, may be more difficult to locate. Also, because the otter's primary means of prey capture in water is by direct pursuit, the improved catchability of certain fish species under ice (see p. 153) may result in their increased representation in the winter diet. This assumes, of course, that the otter is able to gain access to water—a prerequisite that is often problematical.

With the constraints imposed by persistent and often unbroken ice cover, river otters in northern regions are heavily dependent on ponds and streams with steeply banked shorelines that provide denning opportunity with access to water under the ice. Because otter do not excavate burrows themselves, however, this often means reliance on beaver activity and the use of preestablished dens or lodges with their associated plunge holes. Radio-tracking river otter in west-central Idaho over a period of several years revealed that beaver bank dens were used by the otter far more than any other habitat feature.[101] Likewise, in the coniferous forests of Canada, river otter are often found living in close association with beaver, showing a strong preference for beaver-impounded streams as winter habitat. In some cases younger otters, most likely those pushed to the periphery of adult winter territories, have even been observed to deliberately rift beaver dams, dropping water levels behind the dam and creating air spaces under the ice which they may then use for denning. This may also enable them to move through a series of ponds without coming above the ice. So close is the dependency of otter upon beaver activity in boreal regions of Alberta that researchers at the University of Calgary now believe the limit of beaver distribution, the northern treeline, may also establish the range limits of river otter, for lack of suitable winter habitat further north.[102]

Reliance upon anatomical and physiological adaptation, rather than behavioral avoidance of environmental stress, is linked to the usually solitary (in winter) and somewhat nomadic (within territorial bounds) habits of river otter. While they are frequently observed at other times of the year traveling or playing in small groups, most often consisting of the female and her offspring, winter food shortages generally force otters to forage for themselves. Furthermore, in moving about their territories, otters tend to use numerous temporary shelters rather than a central den site. In one case, a single instrumented otter was found to use 88 different resting locations in a 16-month period.[103] Thus, social behaviors like communal denning, grooming, and food sharing, with their attendant energetic advantages, are rare. Conversely, the demands of the river otter's foraging strategy require physiological competence under the most difficult circumstances. Otters must maintain a high level of mobility, dictated by the uneven distribution of prey resources and the undependability of access to water in winter. Centers of foraging activity tend to be widely dispersed within a territory that is shaped largely by drainage patterns. Thus, otters must withstand exposure to low air temperatures in moving overland, while they must also endure and be able to recover quickly from cold-water immersion several times a day. For such challenges the otter shows surprisingly singular adaptation, depending almost entirely on superior insulation.

River otters have a pelage density comparable to the northern fur seal and approximately four times denser than that of the muskrat.[104] The guard hairs are somewhat unusual in that, in addition to providing some insulation through their well-developed medulla, they possess spike-like scales on the hair shaft that interlock, helping to protect the integrity of the underfur. The wavy hairs of the dense underfur in turn form air spaces in both the horizontal and longitudinal plane, which, together with secretions from the sebaceous glands, help prevent water-to-skin contact. While extensive grooming, such as seen with the muskrat, has not been observed among river otter, habits like rolling, shaking the wet fur, piloerection out of the water, and pleating the loose skin all help to restore entrapped air after diving.[105] (It might be noted in passing, that the sea otter, *Enhydra lutris,* possesses the densest hair of any mammal known, fully twice that of the river otter, grooms it meticulously, but lacks arrector pili muscles entirely and thus has no ability to erect the guard hairs.[106] These characteristics reflect a nearly complete adaptation to life in the water for this species.)

The river otter possesses only modest subcutaneous fat, amounting to little more than 10% of its total body weight, with the greatest concentration in the tail. The fact that hair density is also greatest on the tail suggests the importance of insulation in this region, as apparently countercurrent heat exchange mechanisms comparable to those in the tail of the muskrat and beaver are absent.[107] In contrast, the feet of the otter have reduced hair density and little subcutaneous fat, and may serve to exchange heat in those instances (and in southerly locations) where rapid heat dissipation is desirable. During cold-water immersion, or while traveling on the snow, the two large, thin-walled saphenous veins in close association with arteries in the upper limbs may serve as countercurrent heat exchangers to help maintain body temperature. A pronounced slowing of heart rate and increased peripheral vasoconstriction during diving may also reduce heat loss from the feet.[108]

Little is known at present about heat-producing capabilities in this animal. The river otter maintains a basal metabolic rate somewhat higher than predicted on the basis of metabolism/weight relationships for most terrestrial mammals, though this is not atypical of the mustelids.[109] The adaptive advantage of this, if any, is unclear. While an elevated basal metabolism may shift the lower critical temperature of an animal downward slightly (p. 105), postponing the necessity for nonshivering or shivering thermogenesis (pp. 106–7), this advantage would seem to be offset by a higher level of heat production, and, hence, greater energy expenditure throughout the animal's thermoneutral zone. Neither the lower critical temperature of the river otter nor the extent of its seasonal insulative adjustment is presently known.

*Mink (*Mustela vison*)*

Mink share many of the attributes of their larger relative, the river otter, occupying wetlands of almost every description and overlapping in territory with the otter, but minimizing direct competition through dietary habits that are considerably less specialized. Though permanent water appears an absolute requirement in habitat selection by mink, these animals are equally adept land hunters, with birds and small mammals an important constituent of their diet. Figure 73 compares the winter diets of otter and mink from various northern states in the United States.

Just as otter benefit from association with beaver activity, mink often fare better in the vicinity of muskrat colonies, and for much the same reason. The smaller abandoned burrows of muskrat provide ideal den sites for mink in winter, permitting access to both aquatic and terrestrial environments. They have even been known to coexist with muskrat in larger burrow systems.[110] Mink, however, will exploit any prey species that they are capable of catching and killing, including muskrat, and will cache excess food for later use whenever they encounter prey in abundance. Errington has described mink predation on a muskrat colony experiencing a freeze-out, where numerous carcasses were either left in the lodges where they were killed, or dragged to abandoned bank dens nearby. All were consumed in the ensuing month.[111]

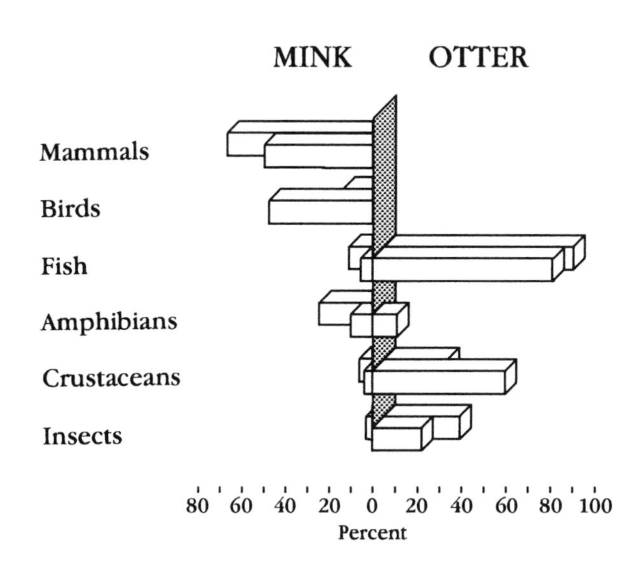

Fig. 73. Frequency (percentage occurrence) of different food items found in scat and gut analyses of mink and otter during winter. Data for mink are from southern Michigan (back row) and Iowa (front row); data for otter are from Minnesota (back row) and Michigan (front row). (Sources: *J. Sealander, "Winter Food Habits of Mink in Southern Michigan,"* Journal of Wildlife Management *7 (1943): 411–17; D. W. Waller, "Feeding Behavior of Mink at Some Iowa Marshes," M. S. thesis, Iowa State University (1962); K. F. Knudsen and J. B. Hale, "Food Habits of Otters in the Great Lakes Region,"* Journal of Wildlife Management *32 (1968): 89–93.)*

The aquatic activity of mink often increases in winter, as difficulty in locating prey species on land may force a shift in diet toward a greater representation of fish. Mink, however, are only marginally adapted to the aquatic environment, and their hunting behavior appears aimed at minimizing time spent in the water. Though mink are capable of finding food while underwater, they are not particularly strong swimmers, having only partially webbed feet and limited lung capacity. Rather, their common mode of "fishing" involves searching along streambanks until they locate vulnerable prey, then plunging into the water after it. Underwater bouts are short, usually of less than 20 seconds duration.[112]

While the mink has a fur density falling between that of the muskrat and otter, it lacks specialized structure such as seen in the latter, and thus, its insulative value does not hold up as well in the water. With immersion, the thermal conductivity of the mink pelt increases 7- to 8-fold over its minimum conductance in air, as compared to a 4-fold increase in the conductance of muskrat fur when submerged.[113] This difference probably reflects the superior air retention characteristics of the highly groomed and oiled muskrat fur. Thus, even with the denser fur and small subcutaneous fat deposits in areas prone to disruption of the fur by swimming, its whole-body insulation in water is quite low and, as a consequence, heat loss quickly exceeds metabolic heat production upon immersion. Only the effectiveness of its peripheral vasoconstriction keeps the mink from cooling at a much faster rate than the muskrat. In one study where mink were sedated to restrict circulatory (vasomotor) response, their cooling rate in water doubled from 0.13° C/min. the same as the muskrat, to 0.26° C/min.[114]

In the balance, mink appear to have sacrificed the specialized adaptations of other semiaquatic mammals for the ability to exploit both terrestrial and aquatic resources effectively in winter. The relatively short fur, paucity of subcutaneous fat, and elongate body typical of this and the smaller mustelids may be maladaptive under cold exposure, but these traits have proven a successful trade for predatory efficiency and flexibility. This seems especially true in winter, when most of the mink's prey species are under snow or ice. It is possible, too, that the energetic disadvantages of the mink's morphology might be compensated for, in part, by selection of well-insulated resting sites and a lowering of resting body temperature, as has been reported for the marten.[115] The winter lower critical temperature (LCT) of mink is unknown, but is likely quite high. The lowest yet reported for any mustelid (the marten) is 16° C,[116] which is considerably higher than even the warmest temperatures northern mink are likely to experience in winter. If the LCT for mink is close to this value, this means that even resting carries high metabolic costs, which may explain the high level of activity for mustelids in general.

GALLINACEOUS BIRDS

Though it might seem an odd comparison, most birds, like the mink, also lead an energetically expensive life-style. Birds generally maintain higher body temperatures than do mammals (40–43° C)[117] and thus tend to have a slightly higher basal metabolism. The small body size of many birds increases their cooling rate (p. 119), while the addition of insulating fat and feathers is limited by both size and the mechanics of flight. Birds also lack brown fat, and therefore must generate sufficient heat to offset losses at low air temperature by shivering almost continuously, whenever they are not generating heat by flight activity (p. 111). This requires nearly constant feeding as daylight hours grow shorter. Hence many birds (not unlike the mink) are almost hyperactive on the coldest days of winter.

Interesting exceptions to these generalizations are found among the gallinaceous ("fowl-like") birds, the several species of grouse and ptarmigan that overwinter in northern forest and tundra regions. While these birds also maintain a body temperature of 40+° C,[118] their larger size, increased insulation, and much reduced activity levels lend some benefits, moderating the high cost of cold exposure. But their single greatest advantage in winter, the one distinctive feature common to all the northern grouse and ptarmigan, is a dietary one. Adapted to feeding on coarse, fibrous plant material, primarily buds, twigs, and conifer needles, these birds have all the advantages of a ruminant without being confined to the ground. Though they face the same the limitations imposed by a high content of cellulose and lignin, as well as plant defense compounds that toxify or reduce the digestibility of plant material (see table 7 and related discussion on p. 168), their compensation is a virtually unlimited source of easily accessible food, requiring little energy to harvest.

The ability of grouse and ptarmigan to handle such a coarse diet is the function of a specialized digestive tract having an enlarged caecum with high concentrations of bacterial flora for the breakdown of crude fiber. Food ground in the gizzard is passed through the small intestine where the more liquid fraction is then diverted into the paired caeca, located at the junction of the small and large intestines. Coarser material bypasses the caecum, moving directly into the large intestine where it is rapidly eliminated.[119] In the willow ptarmigan (*Lagopus lagopus*), a bird that subsists in winter on a large amount of willow twigs, the gizzard action removes little more than the bark and cambium layers from twig segments, leaving the wood essentially intact to be passed undigested.[120] Easily assimilated protein, fats, and soluble carbohydrates are absorbed mainly in the small intestine, while tougher material is left for bacterial fermentation in the caecum. The products of this fermentation include acetic, propionic, butyric and lactic acids, along with ethanol, and may provide a significant portion of the energy requirements of the bird in winter.[121]

As the winter season approaches, the spruce grouse *(Dendragapia canadensis)* of northern forest regions and the blue grouse *(D. obscurus)* of western mountains begin a shift from their eclectic summer diets of berries, grasshoppers, beetles, ants, caterpillars, and the leaves, flowers, and seeds of herbaceous plants[122] to a nearly single-item diet of conifer needles. For the spruce grouse, this change in feeding habit starts long before the arrival of continuous snowcover, undoubtedly requiring gradual adjustments on the part of digestive microorganisms. Though apparently stimulated by early snowfalls, increasing use of trees by spruce grouse is often noticeable in early autumn, and continues through the season even in the absence of snow.[123] Accompanying this shift in diet is a strong drive to seek out sources of grit along gravel roads, lakeshores, streambanks, or uprooted trees, apparently to enhance grinding of the fibrous winter foods in the gizzard. Experienced adult grouse make deliberate, long-distance excursions to familiar sites, congregating in number seemingly as much for the social contact as for the grit, and lingering sometimes for several days. Once snow covers the ground this activity stops and the birds disperse to wintering territories.[124] The ruffed grouse (*Bonasa umbellus*) of mixed broadleaf and coniferous woodlands and the ptarmigan of arctic and alpine tundras likewise shift their diets in fall almost entirely to buds and small twigs of deciduous trees and shrubs. Ptarmigan will continue to gather grit at wind-blown, snow-free sites throughout winter as an aid in the digestion of this woody material.[125]

Accompanying the fall diet change is a gradual increase in the size of the bird's crop, large intestine, and caecum, each contributing to more efficient food collection and nutrient absorption during winter. The combined length of the paired caeca in willow ptarmigan may reach 1.2 m in winter![126] It has often been suggested that the increased fiber content of the diet stimulates winter changes in the gastrointestinal tract, since grouse maintained on commercial poultry feed do not show similar changes,[127] but other studies indicate that with lower ambient temperatures, an increased rate of food intake alone, independent of quality, may result in significant increases in caecal lengths.[128] Regardless of cause, these seasonally adaptive changes underscore the need to process relatively large quantities of food in winter. In addition to higher maintenance energy requirements in the cold, the generally low protein level of winter browse combined with abundant cell wall material and plant defense compounds reduces digestibility,[129] thus requiring additional food intake. The advantage of the enlarged crop and caecum is then twofold, enabling the bird in winter to gather more food, usually in short morning and evening foraging bouts, and to roost for longer periods of the time in a more thermally advantageous location, while digesting.

The ability of northern grouse and ptarmigan to extract energy from nutrient-poor but readily available foods is matched by an equally impressive ability to conserve energy through a combination of physical and behavioral means. The most visible of these

Fig. 74. White-tailed ptarmigan in the alpine tundra, central Rocky Mountains. Note the particularly dense plumage and feathering on the feet. (Photo by Peter Marchand.)

means is the quality of their winter plumage. All are exceptionally heavily feathered in winter, primarily the result of numerous afterfeathers produced from normal body or contour feathers. Afterfeathers are a secondary structure produced from the main axis or rachis of the contour feather, having a plumulaceous (downy) barb and barbule structure greatly enhancing insulation. The afterfeathers of ptarmigan in winter are three-fourths as long as the feathers from which they are derived,[130] adding considerably to the thickness of the body plumage (fig. 74). Similar characteristics are reported for the ruffed grouse in North America and hazel grouse (*Bonasa bonasia*) in Russia,[131] suggesting that this may be typical of the northern grouse in general. The feet of ptarmigan are also densely feathered in winter, which, in addition to insulating them, increases effective surface area four to five times over the nonfeathered foot, decreasing foot loading by an equivalent amount.[132]

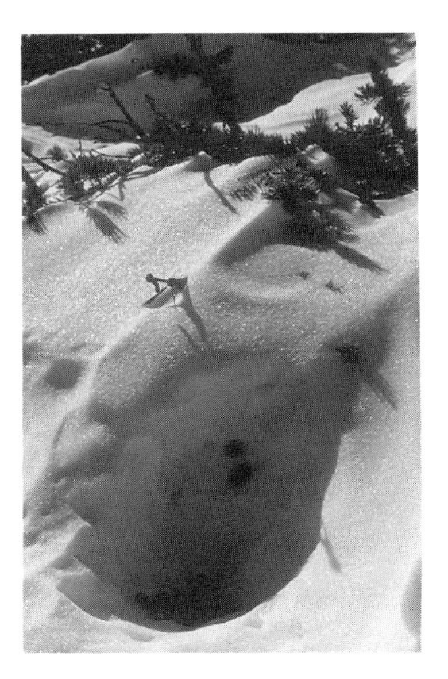

Fig. 75. Roost site of a white-tailed ptarmigan in the alpine tundra. (Photo by Peter Marchand.)

Perhaps the most outstanding aspect of the winter ecology of gallinaceous birds is their habit of roosting beneath the snow (fig. 75). The advantages of this behavior were reviewed briefly in our earlier discussion of thermoregulation in birds (p. 110), where we saw that the energetic savings realized by willow ptarmigan in a snow burrow was rather substantial. Even more impressive are recent calculations by Soviet ecologist A. V. Andreev suggesting that a grouse will spend three times more energy per hour for thermoregulation and related activities on top of the snow than it will within a snow burrow.[133]

While all northern grouse and ptarmigan will roost in the snowpack to some extent, the black grouse (*Lyrurus tetrix*) of northern Eurasia seems to be the most enduring user of snowcover for thermal protection. In Finland, these birds have been reported to spend up to 95% of the winter in snow burrows—essentially all the time that they are not feeding. Within the snowpack they are able to establish, by their own heat-generating activity, a microclimate that closely matches their thermoneutral temperature limits.[134] Shortly after the birds settle into the snow, burrow temperatures begin to rise, concurrent with spasmodic shivering bouts, until the air is several degrees above snow temperature (fig. 76). In some cases burrow temperatures measured close to the bird (but sometimes extending several centimeters into the snow) may rise above 0° C, causing melting and refreezing of the snow within the burrow. The envelope of warmer air thus created, combined with a slight reduction of body temperature (fig. 76), significantly reduces the temperature gradient between the bird and its surroundings, lessening heat loss accordingly. As the temperature rises in the burrow, shivering intensity gradually diminishes, until it ceases altogether when the thermoneutral temperature is reached and birds can thereafter regulate heat loss by insulative adjustments.[135] The lower end of the thermoneutral zone for black grouse is thought to be about +5° C, but this figure is based only on measurements of the onset of shivering. Since the captive grouse in this case did not use their total insulative capacity until temperatures well below that,[136] it is possible that roosting black grouse are comfortable at considerably lower temperatures in the snow.

The blue grouse of montane conifer forests in the western United States and Canada appears to be the antithesis of the Eurasian black grouse. Blue grouse frequently migrate from lower aspen groves to higher conifer stands in winter,[137] where they are able to maintain a positive energy balance, reportedly gain weight, experience little winter mortality, but only occasionally burrow in the snow.[138] Blue grouse prefer to roost in dense subalpine fir or Douglas fir that offer maximum protection from wind and nighttime radiation loss.[139] Excursions away from a roosting site are generally limited to short dawn and dusk flights to nearby conifers where grouse may spend the entire day feeding, preening, and sunning themselves, all with very little energy expenditure. Daily movements in winter average only 70 m.[140]

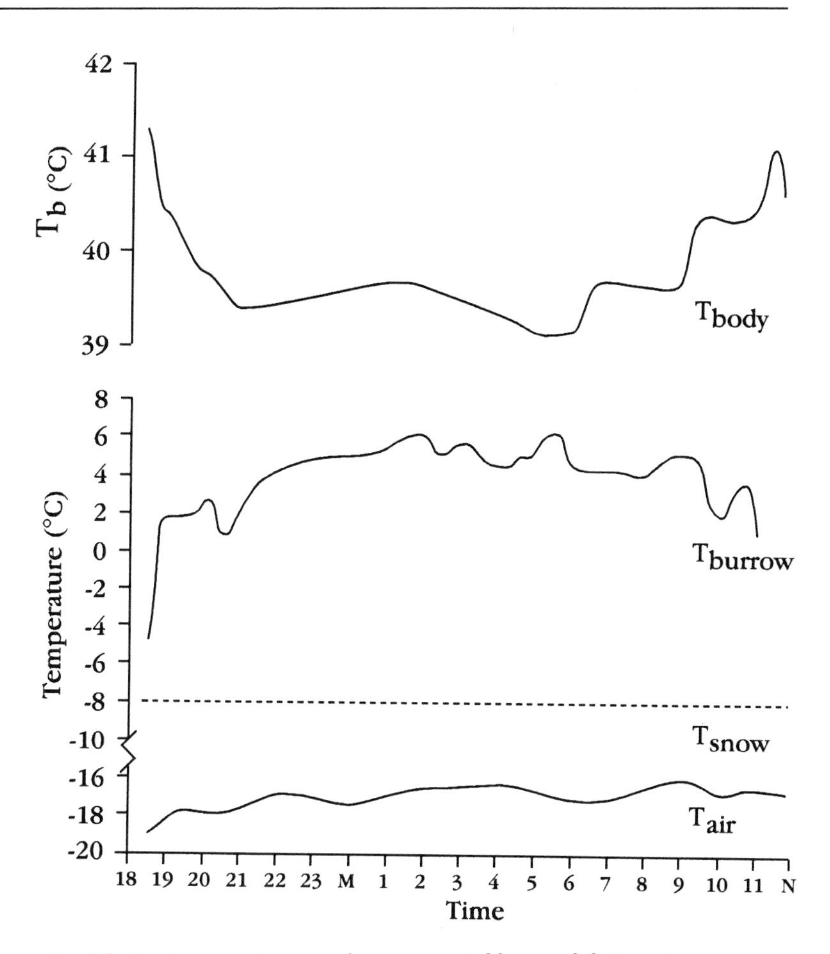

*Fig. 76. Temperatures in a snow burrow occupied by an adult Eurasian capercaillie (*Tetrao urogallus*) while roosting. The captive bird was harnessed with fine wire thermocouples for measurement of body (cloacal) temperature and air/snow temperature 12 mm above the dorsal feathers. It was then allowed to tunnel into the snow in a natural fashion. Temperatures in the burrows of black grouse showed the same general pattern as above. See text for details. (Redrawn from A. Marjakangas, H. Rintamäki, and R. Hissa, "Thermal Responses in the Capercaillie* Tetrao urogallus *and the Black Grouse* Lyrurus tetrix *Roosting in the Snow," fig. 2,* Physiological Zoology *57 (1984): 99–104.)*

Preference of blue grouse for arboreal roosts may relate to superior insulation of these birds. In studying the field metabolic rates of blue grouse in the Wasatch Mountains of northeastern Utah, Peter Pekins and his colleagues found that fasting birds had a lower critical temperature of −5° C, considerably below that of the black grouse and ruffed grouse, both of which roost in the snow. The heat of fermentation (recall our previous discussion under "Northern Cervids") lowered this critical temperature by another 5°, and a drop in air temperature to −20° C resulted in only a 12% increase in metabolic rate. This extraordinary cold tolerance undoubtedly explains why Pekins found field metabolic rates of blue grouse in winter uncorrelated with ambient temperatures in his study area.[141]

With slight variations in physiology and behavior, all of the northern grouse and ptarmigan seem to fare reasonably well in winter. Their low level of activity and selection of optimum roosting sites place minimal energy demands upon them. Daily energy requirements ranging from 420 kJ for the smaller rock ptarmigan (*Lagopus mutus*) and white-tailed ptarmigan (*L. lecurus*) to 600 kJ for blue grouse[142] are easily met in 2-hour morning and evening foraging sessions. Ptarmigan reportedly can subsist for 2 days with a full crop, and the 3 to 5% fat reserves typical of most grouse and ptarmigan can generally subsidize 2 to 3 days of fasting.[143] Only the Svalbard rock ptarmigan (*L. mutus*) and the capercaillie (*Tetrao urogallus*), birds of extreme tundra environments, generally carry more than 5% fat reserves.[144] Thus, with constant forage availability there appears to be little energetic constraint on these birds, as long as there is adequate snowcover to meet their needs for thermal protection. It is in those portions of their range where winter temperatures are low but snowcover less reliable—or frequently crusted, denying their access—that these birds face their most difficult challenges.

8 HUMANS IN COLD PLACES

We have now touched upon the majority of stresses associated with winter and have examined the varied ways in which winter-active organisms deal with the rigors of the season. We have seen many evolutionary options exploited, many different answers to a common set of environmental constraints. Our survey is almost complete. Now it seems appropriate to ask, "What of ourselves?" Humans, warmblooded homeotherms, conspicuous denizens of snow country—what adaptations do we possess for dealing with the winter environment? How well suited physiologically are we to the cold places that we inhabit?

Now and again stories are recorded of humans in cold places showing remarkable tolerance to low temperatures: Australian aborigines sleeping comfortably on the ground, nearly naked, at temperatures close to freezing; Nepalese and Andean mountain people in cold weather wearing only thin clothing without gloves or footwear; Korean women, the Ama, who dive in cold water year-round; Inuit and other northern inhabitants who work with their bare hands at temperatures that would leave most people's fingers numb and useless. These reports notwithstanding, humans for the most part show surprisingly little physiological adaptation to low temperature. The apparent tolerance to cold in many of these northern and mountain peoples is still largely a matter of behavioral or cultural adaptation. Clothing and shelter remain our first line of defense against the cold, and these we utilize with unmatched ingenuity. But take away the covers and we display only modest abilities to control the production and loss of body heat.

By way of natural "covers," the only significant insulative adjustment that the naked human can make physiologically is the addition of subcutaneous fat. It might be suspected, therefore, that humans accustomed to life in cold places would display greater skin-fold thickness (the standard measure for subcutaneous fat) than temperate and tropical beings. Just 1 cm of fat is said to have an insulative value equal to 1 clo,[1] the same as a layer of dry, uncompressed wool or cotton clothing .64 cm in thickness.[2] It has been demonstrated, too, that people having a substantial layer of subcutaneous fat will start shivering at lower air temperatures and when the skin temperature has reached much lower levels than will thin people.[3] This is apparently of decided advantage to English Channel swimmers who, without exception, possess ample body fat.[4] But the fact is that natives of cold climates show a tendency toward leanness instead; in no case has tissue insulation in a cold-adapted race been found to be unusually high. Even the Alacalufe Indians of Tierra del Fuego, who habitually went about nearly naked in one of the coldest, wettest climates exploited by humans, had tissue insulation values comparable to Europeans of similar conformation. The same is true for the Inuit and other arctic peoples, the Andean Indians, and the Korean Ama. It seems somewhat paradoxical that the highest tissue insulation values have been recorded where they are needed least, in well-endowed Caucasians of more temperate latitudes.[5]

CIRCULATORY RESPONSES TO COLD

Discounting, then, the addition of subcutaneous fat as a general adaptation to cold, the most effective physiological mechanism that we have for reducing heat loss when the temperature drops is constriction of blood vessels situated close to the skin surface, and consequent reduction of skin temperature. Vasoconstriction, as it is called, is mediated by sympathetic nerve fibers that release noradrenaline into the vessels in response to signals from cold receptors in the skin or body core. The cold receptors are described as temperature-sensitive "neuronal transducers" whose firing frequency increases as skin temperature is reduced.[6] These relay a message to the hypothalamus, our thermoregulatory control center in the brain, which then orders the noradrenaline release.[7] With constriction of surface vessels, blood is shunted back through deeper veins, reducing heat loss as a result of decreased radiation and conduction from a cooler skin surface (recall discussions on heat exchange in chapter 4). This mechanism is especially effective in the skin of fingers and toes (and in the ears of many other mammals), where vascular channels called "arteriovenous anastomoses" connect directly between the ar-

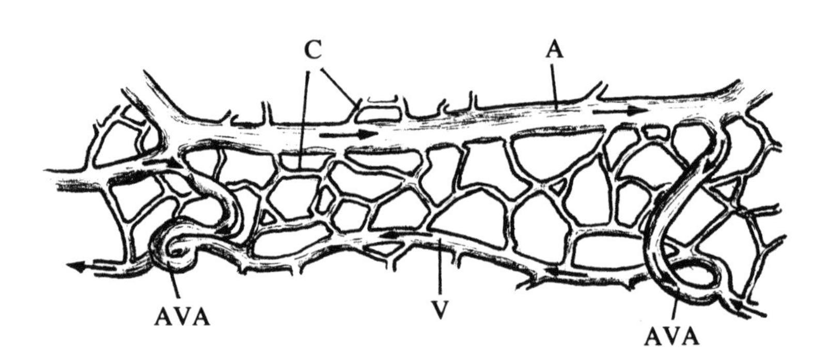

Fig. 77. Capillary beds and circulation in the skin of the extremities. Throughout the superficial layers of the skin are numerous capillary loops that are supplied with blood from arteries and arterioles lying deeper in the dermis. The venules (V) and arterioles (A) of these capillary beds are connected by vascular channels known as "arteriovenous anastomoses" (AVA), which provide a means of bypassing the capillaries (C). During vasoconstriction, the anastomosis opens, arteriolar muscle immediately beyond it contracts, and blood is shunted directly to the venules, thus short-circuiting flow to the downstream capillary network. With vasodilation, the anastomosis closes and circulation is returned to the capillary bed. (Drawn from illustrations and photomicrographs in S. A. Richards, Temperature Regulation (London: Wykeham Publications, 1973), 77, fig. 4.5, and D. I. Abramson, Circulation in the Extremities *(New York: Academic Press, 1967), 117, fig. 26.)*

terioles and venules and provide for a bypassing of the capillary network that lies just under the skin surface (fig. 77). These channels are surrounded by circular smooth muscle with numerous nerve endings under involuntary control. When an anastomosis opens, the arteriolar muscle immediately beyond it contracts, detouring blood through the channel and largely short-circuiting the capillaries.[8]

This is reminiscent of the countercurrent mechanism discussed earlier in animals and, in fact, some authors talk of countercurrent heat exchange in humans. But histological evidence for the presence of a well-developed rete mirabile such as found in the appendages of many animals is lacking. Though a temperature gradient on the order of 10° C from core to extremities is often observed in humans, this is more likely the result of progressive heat loss to the outside by conduction along the length of the limb. In any case, the reduction in temperature of the extremities seen in humans rarely approaches the very steep temperature gradient seen in the appendages of animals with well-developed countercurrent heat exchange systems.

When danger of freezing is imminent, extreme vasoconstriction is followed by short intervals of warming due to intermittent dilation of vessels and closing of the arteriovenous anastomoses, which floods the extremity with warm blood. This response is apparently caused by a rapid loss of ability of the peripheral blood vessels to respond to noradrenaline (the cause of their initial constriction) in extreme cold.[9] The temperature at which vasodilation is first invoked varies greatly among individuals. But once it is initiated, blood flow and skin temperature show more or less regular oscillations (fig. 78) in what is known as the "hunting response" or "Lewis reaction," after the author and his colleagues who first described the phenomenon.[10]

The exact nature of this circulatory response varies among different individuals in an apparently adaptive way and seems to be dependent upon preconditioning—one of the few short-term acclimatory adjustments to cold observed in humans of different races. In one simple and very convincing study, temperate climate subjects were asked to immerse their right index fingers in an ice bath for 20 minutes, four times a day, over a period of one month. In subsequent tests, immersion of the conditioned fingers in ice water resulted in earlier initiation of vasodilation and a more rapid rate of rewarming than in their unconditioned left index fingers.[11] An earlier onset of the hunting response has also been observed in cold-acclimatized Norwegian and Gaspé fishermen.[12] The amplitude of skin-temperature oscillations, however, seems to vary according to the difference between ambient temperature of the body (e.g., under clothing) and that to which the hand is exposed. J. Werner found that as an individual's surroundings were warmed from 15° C

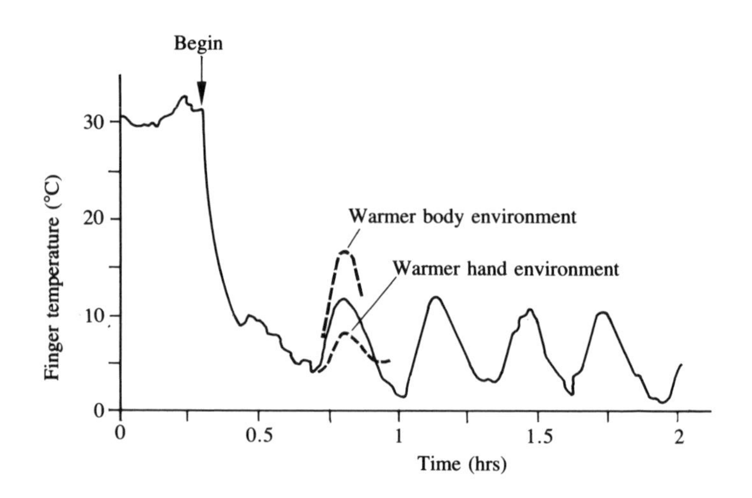

Fig. 78. Recorded "hunting response" in the fingers of a subject sitting in a climatic chamber at 15° C with the hand exposed to an airstream in a cooling box at −5° C. Dashed lines summarize changes in amplitude of skin temperature oscillations when the environmental temperature of either the hand or body is altered relative to the other. (Data are from J. Werner, "Influences of Local and Global Temperature Stimulii on the Lewis-Reaction," Pflugers Archiv. *367 (1977): 291–94.)*

to 30° C, with the hand remaining exposed to freezing temperatures, skin temperature of the fingers showed increasing amplitude of oscillation.[13] Likewise, a decrease in environmental temperature gradient between the body and hand reduced the amplitude of oscillations (fig. 78). Interestingly, though, when test subjects were put under heat load, at an ambient temperature of 45° C, the hunting response in the hand exposed to freezing temperatures was completely cancelled.[14] These results suggest an integration in the hypothalamus of signals sent from receptors in peripheral tissues and the body core.

Curiously, in some northern peoples it seems that early and prolonged vasodilation in the extremities, rather than severe constriction of surface vessels, is the normal response to cold (an extreme modification of the hunting response, perhaps?). The fingers of

Inuit adults, when immersed in ice water, warm after only brief exposure and may be maintained 4° to 5° higher on average than the fingers of other subjects unaccustomed to the cold.[15] The warmer hands of Inuit men can be maintained for relatively long periods of exposure to cold air, the difference in skin temperature between Inuits and outdoor whites being at times as much as 10° when the fingers of whites are undergoing extreme vasoconstriction (fig. 79*a*).[16] Accompanying this response is a pronounced difference in heart rate and blood pressure, both being very much lower in the cold-acclimatized Inuit (fig. 79*b*).[17]

The increase in blood flow with vasodilation (which can reach as high as 100 times the minimum flow to superficial tissues!)[18] greatly accelerates heat loss, of course, but has the adaptive advantage of increasing manual dexterity at low temperatures. Apparently the increased heat loss with vasodilation can be afforded in the case of the Inuit when other mechanisms (i.e., cultural) for conserving heat are so effective. There is some reason to believe, too, that the increased flow of blood in the hands of these and other people such as Norwegian fishermen may be conditioned by physical exercise, especially at low temperatures, since the same can be influenced by fitness training.[19]

The mechanism for such acclimatory response has been difficult to isolate. It has been suggested that long-term cold exposure may result in reduced sensitivity to the vasoconstrictor effects of noradrenaline, an effect that can be demonstrated experimentally. The rise in systolic blood pressure that accompanies the shock of cold immersion and vasoconstriction in nonacclimatized subjects can be induced by intravenous injection of noradrenaline; but when noradrenaline is administered to subjects following repeated cold exposure, its effects are diminished.[20] In a case study of Quebec City mailmen, blood pressure and heart rate were found to be substantially lower at the end of winter, as compared to the beginning of winter.[21] It is possible that some change in the quality of skin following prolonged cold exposure—a change, for example, in moisture content or thickness of the outer, nonvascular layer of the epi-

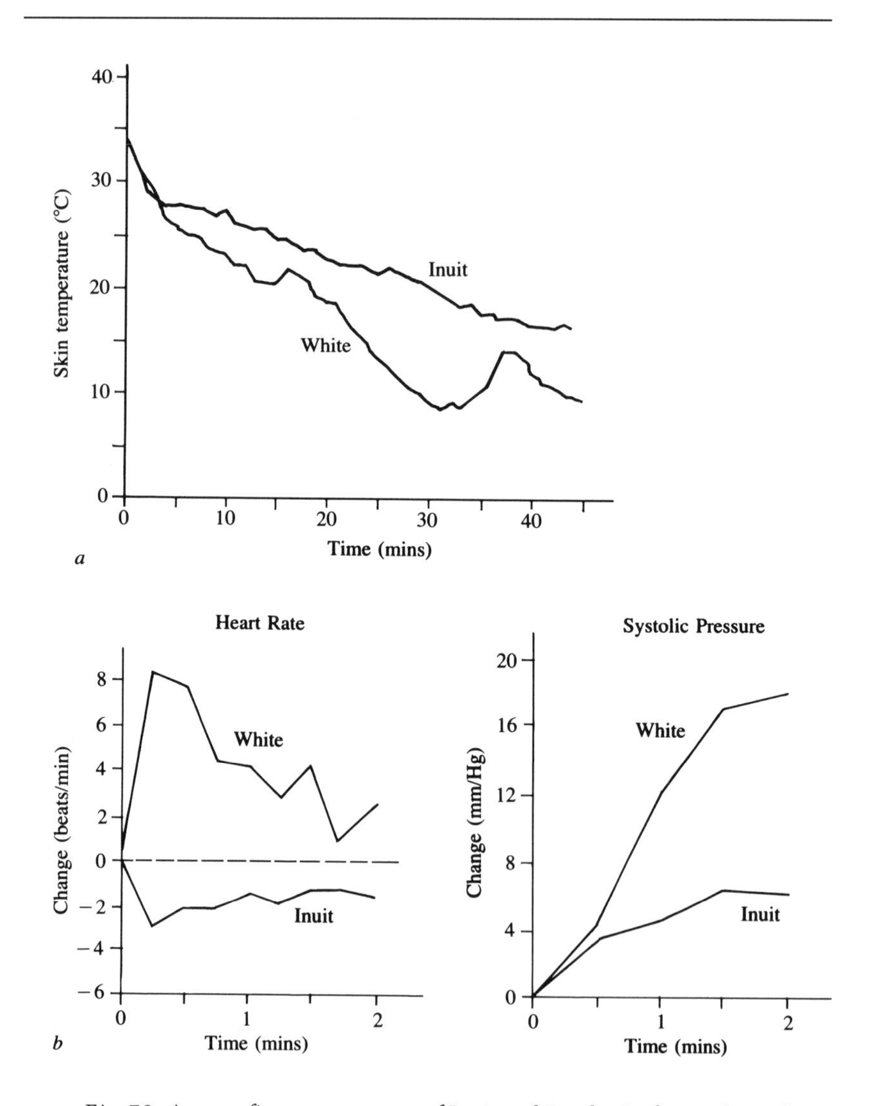

Fig. 79. Average finger temperatures of Inuit and "outdoor" white males with hands exposed to air at −3° to −7° C (a) *and comparative changes in heart rate and systolic blood pressure with immersion of the hand in cold water* (b)*. (Redrawn from L. K. Miller and L. Irving, "Local Reactions to Air Cooling in an Eskimo Population,"* Journal of Applied Physiology *17 (1962): 449–55* (a) *and J. LeBlanc, "Adaptation of Man to Cold," in L. C. H. Wang and J. W. Hudson, eds.,* Strategies in Cold: Natural Torpidity and Thermogenesis (New York: Academic Press, 1978), 695–715 (b).)

dermis—might alter the thermal conductivity of the skin and, hence, the actual temperature of the peripheral cold receptors.[22] Alternatively, a change in blood flow during cold acclimatization might lead to an overall increase in hypothalamic temperature, with this, in turn, causing a change in perception of information arriving from the periphery.[23]

INCREASING HEAT OUTPUT

On the heat production side of the equation, humans fare a little better, but still lack some of the regulatory mechanisms exhibited by other mammals of cold-winter climates. Increased muscle activity, either voluntarily or through shivering, remains out principal mechanism for increasing heat production. The skeletal muscle that makes up roughly 50% of our body mass contributes about 20% of our total heat production while at rest, but vigorous exercise can increase heat output as much as 10-fold. Shivering by itself can raise the metabolic rate fivefold.[24] This response, as with vasodilation, is also triggered by cold receptors in the skin that transmit nerve signals to the hypothalamus. Maximum shivering response seems to occur at a skin temperature of 20° C; but even before shivering, a measurable increase in muscle tone occurs, often felt as a tightening of the neck and shoulder muscles, and this alone can double heat production.[25]

From an energetic standpoint the efficiency of shivering is low, perhaps only 10% of the energy invested actually warming the core. This is primarily because of enhanced heat dissipation through increased blood flow and increased convective heat loss associated with rapid movements.[26] From a practical standpoint shivering is mentally disturbing and reduces functional efficiency. For these reasons, *decreased* shivering at low temperatures is usually taken as an indication of cold adaptation in humans. The implication here, however, is that shivering thermogenesis is replaced in the cold-acclimatized individual by some form of nonshivering thermogenesis, and this is problematical.

In a review of the literature on native peoples of cold regions, including the Inuit Eskimo, Athapaskan and Alacalufe Indians, Australian aborigines, Kalahari Bushmen, Korean Ama, Norwegian Lapps, and the Peruvian Quechaus, there is little indication of nonshivering thermogenesis in any.[27] However, in the case of the Alacalufe Indians, there may be room for question. As already noted, these people inhabit one of the more inhospitable places in southern Argentina and Chile, where nighttime temperatures are never far from freezing, often accompanied by rain and sleet, and where daytime temperatures may not warm much above 10° C. Yet the Alacalufe have traditionally slept naked in rudimentary shelters without shivering and without any drop in body temperature during the night. Apparently, they maintain a stable body temperature and remain comfortable under these conditions by means of an elevated metabolic rate—some 30 to 40% higher than in other peoples.[28] This is certainly suggestive of effective nonshivering thermogenesis, though the possibility has not yet been tested.

Other evidence suggests that with extended exposure to cold a nonshivering response may be elicited in humans wherein basal heat production rises in relation to increased glandular activity. Studies of acclimatization among expedition members to Antarctica during their first year of stay revealed steadily increasing body-core temperatures when subjected to standardized cold-exposure tests.[29] It appears that hormones secreted by the thyroid, pituitary, and adrenal glands may somehow be responsible for such a prolonged increase in metabolic rate.[30] One might reasonably speculate that an increased metabolism of brown fat reserves, mediated by one of these hormones, is the actual mechanism for the elevated heat production (as was the case in other mammals), but reports of brown fat in human adults are contradictory. In analyzing a number of whole-body thermograms, researchers found no evidence suggesting the presence of metabolically active brown fat in adults.[31] Others report localized deposits around the kidneys and in the aortic region.[32] While it is tempting to suggest that a reversion of white fat to brown might occur in adults under cold stress, as has been seen in laboratory rats, evidence here is also scant. However, histological and biochemical analyses from a number of Finnish outdoor workers (obtained posthumously) did reveal thermogenically active fat deposits around the neck arteries, leading researchers to conclude that working in the cold could result in the retention (or production) of brown fat in "strategic" places in human adults.[33]

Considering all the evidence available to date, it appears that only in infants, where brown fat may comprise 1 to 6% of their body weight, is this mechanism for nonshivering thermogenesis significant in the maintenance of body temperature.[34] The most likely explanation for the elevated heat production observed in cold-acclimatized adults, in the absence of shivering, is that cold stimulation of hormonal activity promotes the conversion of glycogen to glucose (mediated by adrenaline) and increases the rate of oxidative metabolism (mediated by thyroxine), resulting in increased heat output by the liver (fig. 80).[35] It should be emphasized, however, that non-shivering thermogenesis in humans, excepting perhaps the case of the Alacalufe Indians, appears to be of minor importance when compared to the role of muscle activity in increasing heat production under cold stress.

As we saw in chapter 4, one alternative to boosting heat production at low ambient temperatures is to reduce body core temperature. This is an uncommon physiological strategy among humans living in cold places, but there are two outstanding examples of it among populations exposed nocturnally to low temperatures. These are the Australian aborigines and the Kalahari Bushmen, both of whom customarily sleep on the ground at night nearly naked, with temperatures often only a few degrees above freezing. Unlike the Alacalufe Indians, however, these people undergo nightly torpor (actually a shallow hypothermia), in which they suppress shivering, reduce heat production, and allow their body temperature to drop through the night, while they sleep in apparent comfort. This provides a dramatic contrast with a group of European males en-

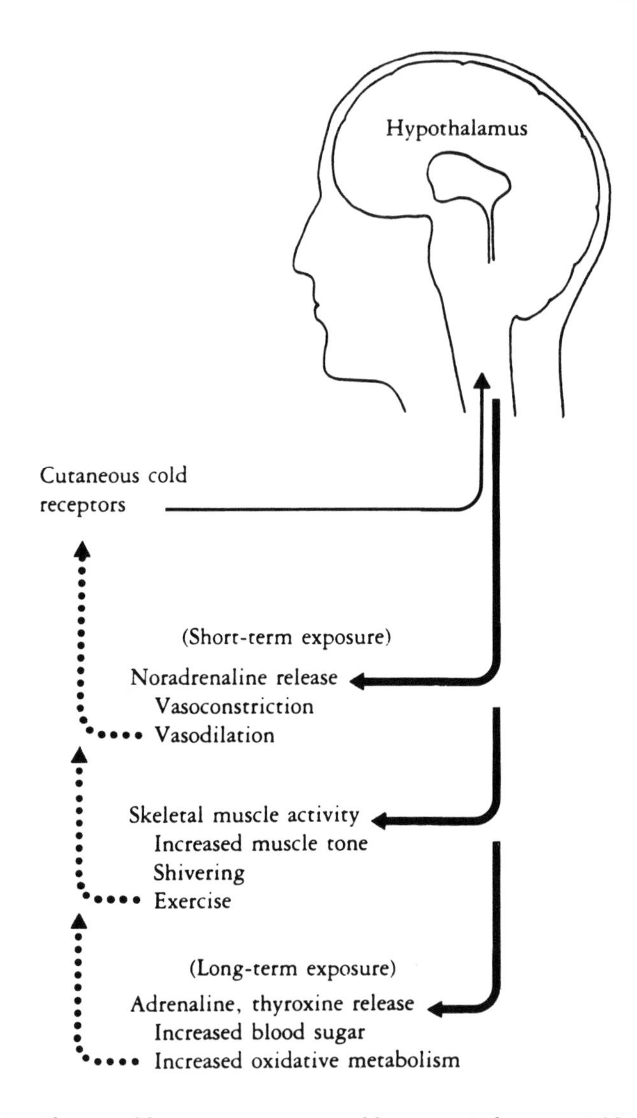

Fig. 80. Short- and long-term responses to cold exposure in humans. Cold receptors in the skin transmit nerve signals to the hypothalamus, our thermoregulatory control center, which then mediates the responses indicated. Increased heat production feeds back through skin receptors (dotted line) to regulate the input signal. (Redrawn from A. Holdcroft, Body Temperature Control in Anesthesia, Surgery and Intensive Care *(London: Bailliere Tindall, 1980).)*

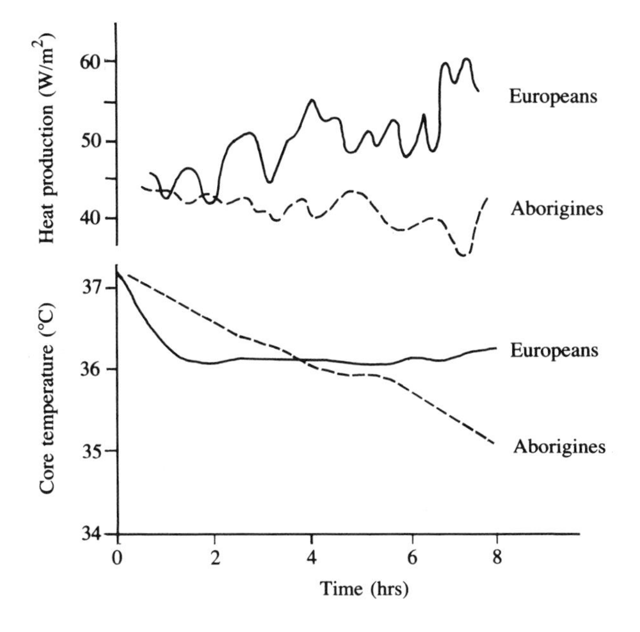

Fig. 81. Heat production and body temperature in a group of aborigines and Europeans during a night of moderate cold exposure. The shallow hypothermia of the aborigines is an uncommon mechanism in humans for reducing heat loss and maintaining body temperature, of adaptive advantage only where daytime warming in their native environment is certain. (Redrawn from S. A. Richards, Temperature Regulation *(London: Wykeham Publications, 1973).)*

camped with the aborigines and also attempting to sleep naked while exposed to moderate cold. The Europeans maintained a constant body temperature, but only by the increased heat output that came with shivering uncomfortably through the night (fig. 81).[36] The nightly hypothermia displayed by the aborigines and Bushmen is of adaptive advantage, it seems, only in an environment where daytime warming is assured; a similar physiological response might put mountain and northern people at great risk. With repeated exposure, whites have been able to lower their shivering threshold and reduce their level of discomfort, so that sleeping in the cold becomes possible, but only with continued high metabolic expense.[37]

MIND OVER TEMPERATURE: PSYCHOLOGICAL RESPONSES TO COLD

While it is possible to demonstrate some improved cold tolerance with physical conditioning, it is much more difficult to assess the role of psychological preparation in dealing with the cold, though such is likely to play a significant role in the comfort of "foreigners" in cold places. Lucy Kavaler, in her popular book *Freezing Point,* relates a story of how military personnel tried to develop a set of physiological criteria by which they might determine who among new recruits would be best suited for duty in the North and who should be sent to the tropical zone. It turned out in the end that the best test of a soldier's ability to adjust was to ask each whether they liked cold better than heat.[38] Their stated preference might, of course, be related to some fundamental difference in physical makeup; but nonetheless, there seems no substitute for mental acceptance of the cold.

As a case in point, Lawrence Irving, who has studied numerous human subjects in the North, once noticed a group of white students in Fairbanks, Alaska, who, for reasons of their religious cult, adopted the custom of going about barefoot and lightly clothed in winter, apparently without experiencing great discomfort. When the students consented to testing, Irving found that they indeed had considerably greater tolerance for cold than a control group of recruits from the local air force base.[39] Voluntary acceptance of cold exposure likely had much to do with their tolerance, but Irving noted also that during the cold tests the students sat quietly studying for an exam while the airmen shivered uncontrollably. This further suggests some mental control over perceived comfort levels. In more recent testing at the Hypothermia Laboratory in Duluth, Minesota, volunteer subjects shivering violently in a cold-water tank stopped shivering altogether when required by researchers to concentrate on simple mental arithmetic—adding, subtracting, and dividing a series of numbers, for example.[40]

ADAPTIVE RESPONSES OF PEOPLES HABITUALLY EXPOSED TO COLD

Humans throughout the world have for the most part relied on cultural adaptations for their comfort in cold places. Some groups, however, particularly those of nontechnological cultures, have evolved subtle but effective physiological mechanisms for dealing with the cold. Still others, habitually exposed to cold in outdoor occupations, show varying degrees of short-term acclimatization. Summarized below are a number of human adaptations to cold referenced in this chapter:

Group	*Response to cold exposure*
Australian aborigines	Shallow nighttime hypothermia; suppression of shivering
Alacalufe Indians	Elevated metabolism*
Inuit Eskimo	Increased circulation and skin temperature in extremities; reduced heart rate and blood pressure
Norwegian and Gaspé fishermen	Increased circulation and skin temperature in extremities
Quebec City mailmen	Reduced heart rate and blood pressure
Antarctic workers	Increased core temperature*
Finnish outdoorsmen	Thermogenically active brown fat in "strategic" locations*
Tibetan and Indian Yogis	Increased temperature of extremities and core; delayed shivering response*

**Possible presence of nonshivering thermogenesis*

It has long been known, too, that yoga training, whether due to its physical or meditative benefits or both, confers some advantage in increasing cold tolerance. A group of Tibetan Buddhists who live in unheated, uninsulated stone huts in the Himalayan foothills and who practice an advanced form of meditation known as g Tum-mo yoga, show an extraordinary ability to elevate skin temperature in their extremities by as much as 8° within an hour of assuming their meditative posture.[41] Similar advantages of yoga have been demonstrated experimentally in studies conducted by the Defense Institute of Physiology and Allied Sciences in India. In one such experiment, two groups of randomly selected army recruits were prescribed two different training regimens, one a routine physical-exercise program emphasizing endurance and muscle strength, and the other a yogic exercise program of equal duration that emphasized relaxation and controlled breathing. At the end of six months, the recruits were tested for physiological response to cold exposure. Sitting naked in a room at 10° C for two hours, the recruits who had completed the yoga training were found to maintain higher core temperatures with a much-delayed shivering response, as compared to the physical-training group.[42]

It is clear that with sufficient acclimatization, humans are, by one means or another and in varying degrees, able to increase their tolerance of the cold (see box of "Adaptive Responses of Peoples Habitually Exposed to Cold," p. 253). But as Irving has noted, it is usually only under the most pressing of circumstances that the human species accepts the degree of prolonged exposure necessary to acclimatize successfully to the cold. Instead, most individuals go to great lengths to avoid such exposure. As often expressed by the Inuit, the best way in the end to deal with cold weather is to "take care not to be cold."[43] By our cultural and technological ingenuity, we have exploited the coldest places on earth. Biologically, however, we remain essentially tropical beings.

APPENDIX A

Measuring Temperatures of Microhabitats and Small Objects: Thermocouple Thermometry

When probing the temperatures of winter microhabitats like insect galls or vole nests, or monitoring freezing processes in plants, insects or other organisms, it is important to choose a sensing device of very small mass and, in many cases, of remote measuring capability. This is necessary so that the temperature of the object in question is not affected by the act of obtaining the data. Thermocouple thermometry offers such advantages.

Thermocouples are simple temperature sensors formed from the junction of two wires of different composition. Copper and constantan are the materials most commonly used for ecological studies. In practice, the wire (often manufactured in duplex strands) is cut to the desired length and one pair of ends stripped, twisted, and soldered to form a probe (fig. A-1). The other ends are connected, usually via a small male plug, to a thermocouple potentiometer. This instrument has a built-in electronic temperature reference (normally 0° C), against which the temperature of the test probe is compared. Whenever the two are at different temperatures, a voltage is generated (measured in millivolts) that is proportional to the temperature difference between them. Accuracy is usually within 1% of this temperature difference.

Thermocouple probes can be custom-made for appropriate length and size quickly and relatively inexpensively. They are sensitive to tenths of a degree and equilibrate in a matter of seconds with the test subject, allowing rapid temperature measurement with a minimum of operator influence. Copper-constantan wire is readily available in a range of diameters down to the thickness of a spider web and, thus, can be used with very small objects. The internal tem-

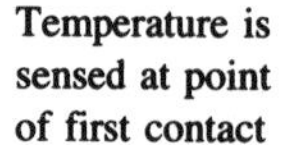

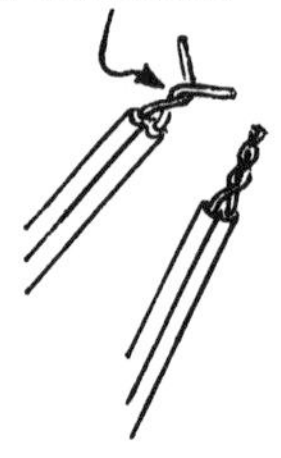

Fig. A-1. Thermocouple junction.

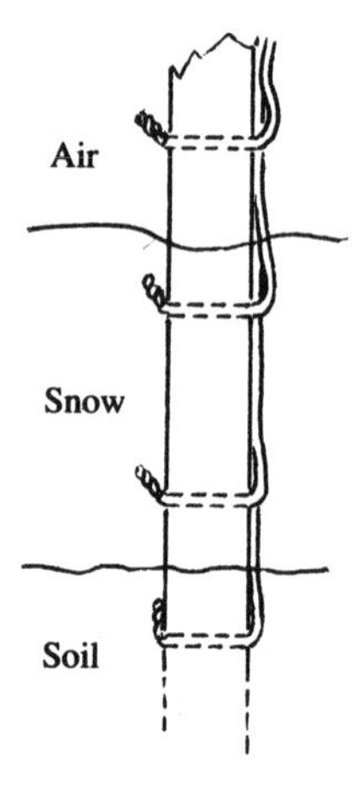

Fig. A-2. Standard for obtaining temperature profiles.

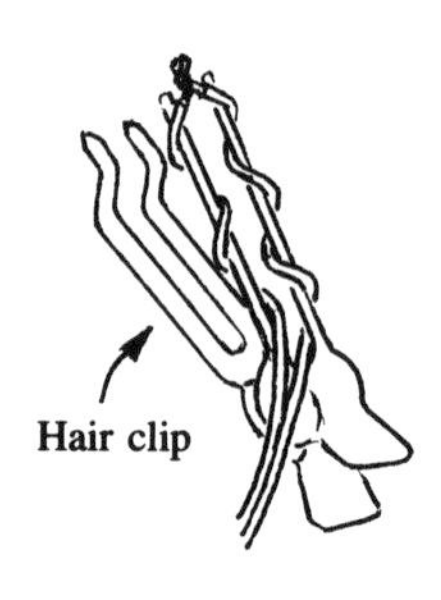

Fig. A-3. Field probe for measuring plant temperatures.

peratures of conifer needles reported in chapter 3, figure 21, were measured in this way.

Temperature profiles in soil, snow, and air above the snowpack can be obtained easily by drilling a dowel at selected intervals, pushing the thermocouple wire through the holes and turning the probe ends up, and then driving the dowel into the ground to the desired depth before freeze-up (fig. A-2). A convenient field probe for measuring leaf temperatures can be made by modifying a spring hairclip, separating the copper and constantan wires and wrapping each around opposite prongs of one leg, and forming the thermocouple junction in the space between them (fig. A-3).

Instruments and materials for thermocouple thermometry may be obtained from:

Omega Engineering Inc.
P.O. Box 4047
Stamford, CT 06907–0047
(203) 359–1660

For general use in measuring soil, snowpack, and air temperatures, characterizing nest or den microclimates, or for studies with woody plants, 24-gauge wire is useful and durable. For smaller objects, 30-gauge wire may prove more versatile.

APPENDIX B

Measuring Insect and Plant Freezing Resistance: Differential Thermal Analysis

Differential thermal analysis (DTA) is a technique commonly used for measuring freezing events in living (hydrated) plant and animal tissues. The procedure involves simultaneous measurement of air temperature and the temperature of the test subject while both are being cooled slowly. The freezing point of the subject is detected by a temporary increase in tissue temperature, relative to air temperature, as latent heat is released. This appears as an exotherm on a continuous plot of tissue-minus-air-temperature difference versus air temperature in the graph below. Generally speaking, in plants acclimated to winter conditions, cell injury does not occur at the first freezing exotherm, which represents extracellular ice formation, but rather tissue death occurs at a second and much lower exotherm.

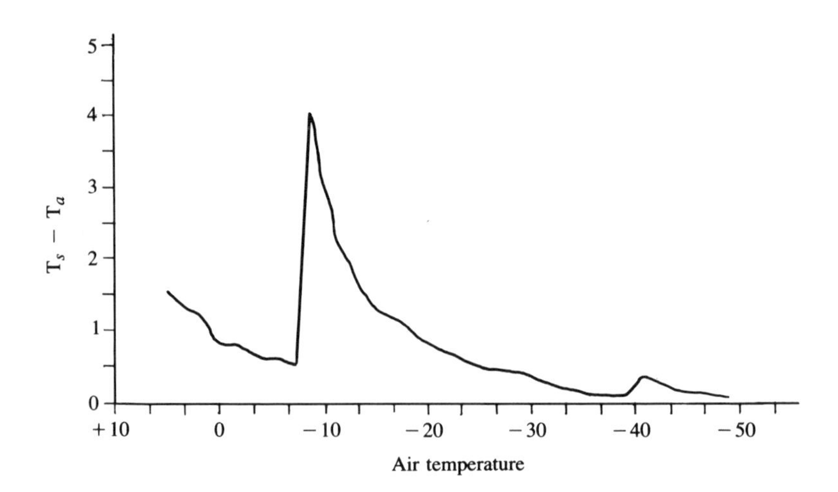

Insects, on the other hand, will display only one exotherm, the temperature at which it occurs differing markedly between freeze-tolerant and freeze-susceptible insects.

In performing these analyses it is important that the rate of cooling be carefully controlled (on the order of 1° C/min) so that an artificial killing point does not result from premature intracellular freezing. Described here is a simple and low-cost method for obtaining this objective utilizing a dry-ice/methanol-bath setup that will typically bring plant tissue temperature down to −60° C in approximately an hour and a half. A step-by-step outline for measuring freezing resistance in woody plant material follows:

1. Cut stem or tissue section to be tested approximately 5 cm in length.
2. Carefully puncture the bark or epidermis with a needle at an angle of about 20° from the axis of the stem. Insert a thermocouple probe so that the twisted wire is *entirely* under the surface and hold it in place using thin strips of masking tape as shown in the diagram below.
3. Mount a second thermocouple to the outside of the specimen with tape to serve as an air temperature probe, and position the tip so it is about 2 to 3 mm away from the surface and lower on the specimen than the first thermocouple. An alternative procedure is to imbed the second thermocouple in an oven-dried stem section placed adjacent to the test specimen, to serve as a control.
4. Place a wad of cotton around the end of the test subject to prevent contact with the sides of its container. Insert the specimen with its thermocouple wires into a centrifuge tube and place the whole

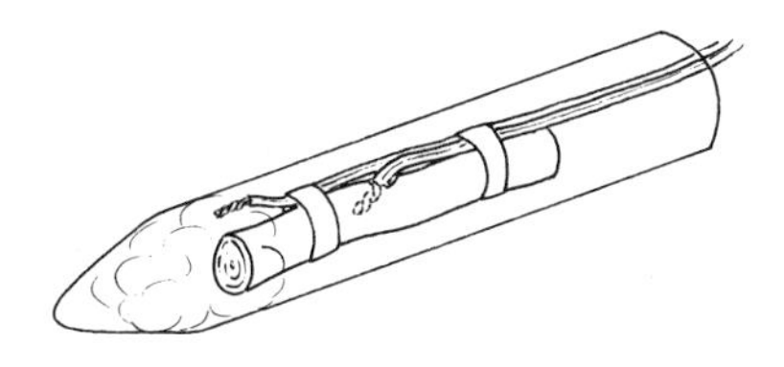

assembly in a flask containing 60% methanol (prechilled to about 5° C), being careful not to spill the methanol into the tube containing the specimen.

5. Stopper the flask, place it in an insulated box and pack it loosely in dry ice. Connect the thermocouples to a potentiometer (a switch panel is useful for multiple specimens) and begin recording air and stem temperatures at 60-second intervals.
6. Plot tissue-minus-air-temperature differentials ($T_s - T_a$) as a function of air temperature.

If you have been careful in the preparation of your specimen, your graph should look somewhat like figure B-1 below, and you should have no trouble identifying the freezing exotherms. Note, however, that the freezing resistance of some of our hardiest plant species, when fully acclimated, may exceed the limits of our experimental temperatures and, thus, the second exotherm may not appear.

Freezing curves for insects are obtained in the same manner as described above, except that the thermocouple probe for the test subject may simply be placed in contact with the specimen and the two (insect and probe) rolled together in cotton. The air-temperature probe is then placed on the outside of the cotton ball. Results are plotted as for the plant samples.

APPENDIX C

Measuring Respiratory Oxygen Consumption of Small Invertebrates

The standard approach to determining metabolic rates of animals is to measure their rates of oxygen consumption while fasting and at rest. This is usually accomplished by placing the animal in a closed chamber having a flow-through air supply in which oxygen levels can be measured with a gas analyzer on both the intake and outflow sides. While this is a relatively simple and accurate (though not inexpensive) means of obtaining metabolic rates for small mammals and birds, the volume of air flow required is far too great for measuring oxygen consumption of very small organisms like insects. For these, a simple manometric technique is used instead. Described below is a microrespirometer based on such a technique and developed by R. E. Lee, Jr. for studies of insect metabolism at low temperature. It can be assembled quickly and inexpensively with readily available materials, and can be adapted easily for use with small aquatic invertebrates.

In principle, the organism is placed inside a very small enclosure in which it respires, consuming oxygen and liberating carbon dioxide in a 1:1 ratio. Through use of a chemical absorbant, such as potassium hydroxide (KOH), to remove the expired CO_2 from the chamber atmosphere, a partial vacuum is created within the chamber as oxygen continues to be withdrawn by the organism. The rate of oxygen consumption can then be measured by following the movement of an indicator drop in an open-ended capillary tube (the manometer) that is connected to the chamber. The indicator drop is drawn down the manometer by the progressive increase in partial vacuum as oxygen is depleted.

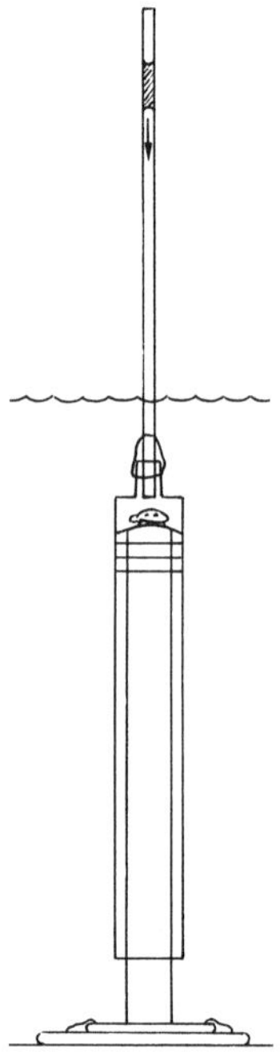

The respirometer chamber used here is a small 1, 2, or 5 cc plastic syringe (choice of size depending on the organism being studied). To the tip of the syringe, a disposable glass micropipette (10 or 20 ul) is attached using thermoplastic glue. This serves as a manometer. The end of the syringe plunger is then weighted with a metal washer so that the whole apparatus can be positioned upright in a constant-temperature water bath as shown in the accompanying figure. In use, the test subject is introduced into the barrel of the syringe, the plunger inserted until the organism is confined at the end of the syringe, and the respirometer then placed in a water bath such that the syringe is completely submersed, with only the micropipette extending above the water. Small aquatic invertebrates can be placed in a water-filled microcentrifuge tube or similar vessel and loaded upright into the syringe in the same manner.

After placing the respirometer in the water bath and allowing 10 to 15 minutes for temperature equilibration (very important!), a small drop of 10% KOH is introduced to the open end of the micropipette. This serves as both the CO_2 absorbant and the manometric indicator fluid. At 10- to 20-minute intervals thereafter, the position of the KOH droplet is measured with respect to an initial reference point, until a steady rate of uptake is observed (it is especially important that measurements are made without handling the respirometer, as very small temperature changes can affect results significantly). An empty respirometer situated in the water bath and serving as a thermobarometer is monitored simultaneously to correct for movement of the indicator drop resulting from changes in water temperature or barometric pressure.

Oxygen consumption is calculated by multiplying the distance moved by the KOH droplet (mm/unit time), averaged over the period of steady uptake, by the volume of the micropipette used (ul/mm). Results, in microliters of oxygen per unit time, are usually expressed on a weight or "per individual" basis. A series of measurements conducted over time and a range of temperatures then gives a good picture of metabolic compensation and acclimation rates in overwintering insects.

APPENDIX D

Measuring Dietary or Habitat Preferences in Winter:
The Rank Preference Index

Many wildlife managers believe that an animal's winter condition is reflected in its feeding behavior and that optimal food selection may determine, in large part, its overwintering success. This might be especially true of browsing animals, where the diversity of food resources above a deep snowpack is often limited and where the chemical quality of browse species may play a major role in dietary preference. Thus, an analysis of food utilization relative to availability could prove a valuable exercise in the study of animal energetics in winter.

The following method, modified slightly from D. H. Johnson,[a] is used to assign a rank-order preference to plant species browsed by animals like deer and snowshoe hares that leave clear feeding "signatures." The method involves separately ranking the availability of browse species and then the utilization of those species, with the difference in rank between use and availability for each species providing a measure of relative preference. The field procedure is as follows:

1. Define a feeding site in which there has been obvious and relatively recent browsing activity. It is usually clear from tracks, scat, or other signs which animal species have been feeding. (When fresh snow obliterates all other signs, remember that twig ends clipped by snowshoe hares are cleanly chiseled with their sharp upper and lower incisors, while twigs browsed by deer, including elk and moose, tend to be ragged from being ripped off rather than cut cleanly.) The boundaries of the feeding site, as defined for purposes of listing plant species, are usually sub-

jective and based on visible signs or field experience on the part of the observer. In most cases, the "limit of sight" while standing at the center of the feeding area will be sufficient to define the boundaries.

2. List all of the available plant species within the feeding area and then assign a rank to each, beginning with "1" for most available. A sample data card is illustrated below (fig. D-1). In the case of "ties," an average rank may be assigned. Keep in mind that "available" means having within reach those plant parts that are normally eaten by the animal. A tree, for example, with its lowest branches 2 m above the ground may be available to a deer, but should not be listed as "available" in a snowshoe hare browse survey, unless it is established that snowshoe hares eat the lower bark of that species.

SNOWSHOE HARE FEEDING PREFERENCE

Date: 1/12/91

Location: BRAINARD LAKE, CO.

	Rank Order (1 = most)		
Species Present	Usage	Avail.	Use-Avail.
LODGEPOLE PINE	1	1	0
ENGELMANN SPRUCE	Ø	2	5
SUBALPINE FIR	4	3.5	0.5
TREMBLING ASPEN	3	3.5	-0.5
COMMON JUNIPER	2	5	-3

Fig. D-1. Sample data card.

3. Now evaluate browse signs and rank the available plant species according to their usage, again assigning a "1" for most browsed.
4. Subtract the availability rank from the usage rank for each plant species on the data card (differences may be positive, zero, or negative). Next, average the rank differences for a given plant species over all the feeding sites surveyed and order all species according to average rank differences. This will show relative feeding preference. In essence, those plant species having an average rank difference near zero are browsed roughly in proportion to their availability; those having a negative difference tend to be preferred (browsed more than expected on the basis of availability); and those having a positive difference tend to be avoided (browsed less than expected on the basis of availability). This is summarized in the following diagram:

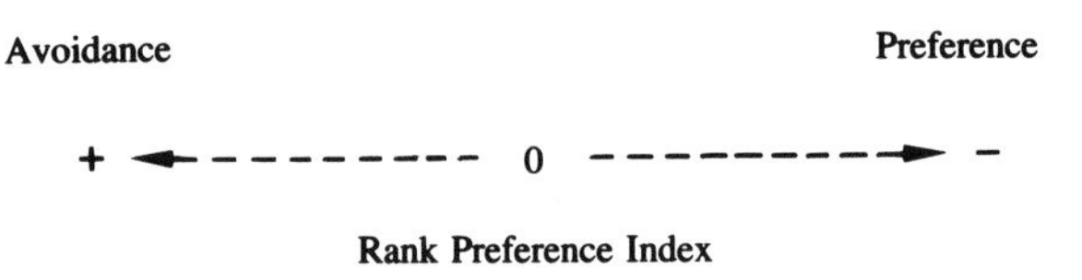

One of the principal advantages of Johnson's rank-preference index lies in the fact that inclusion of common but seldom eaten food items in the analysis does not affect the outcome—a common fault of other methods. Note, however, that there is an important distinction between "least used" and "not used." Unbrowsed plant species may be assigned a zero in the rank-order column, but then must be assigned an arbitrary (i.e., "dummy") value for rank difference equal to the total number of species on the survey card in order to avoid artificially high preference ranking in certain cases. The procedure outlined above may be used in exactly the same manner to rank habitat preferences of animals by substituting habitat types for browse species.[b]

[a]D. H. Johnson, "The Comparison of Usage and Availability Measurements for Evaluating Resource Preference," *Ecology* 61 (1980):65–71.

[b]An example of such application is illustrated in Johnson and reproduced in C. J. Krebs, *Ecological Methodology* (New York: Harper Row, 1989).

NOTES

Chapter 1. Winter Paths

1. K. Schmidt-Nielsen, *Animal Physiology: Adaptation and Environment.* 3rd ed. (New York: Cambridge University Press, 1983), 181.
2. Ibid., 215.
3. B. Barnes and D. Ritter, "Patterns of Body Temperature Change in Hibernating Arctic Ground Squirrels," in C. Carey, G. Florant, B. Wunder, and B. Horwitz, eds., *Life in the Cold: Ecological, Physiological, and Molecular Mechanisms* (Boulder, CO: Westview Press, 1993), 119–30.
4. B. Barnes, C. Omtzigt, and S. Daan, "Hibernators Periodically Arouse in Order to Sleep," in Carey et al. *Life in the Cold,* p. 164.
5. See literature review in M. Kawamichi and T. Kawamichi, "Hibernation Commencement and Emergence in Chipmunks," in Carey et al., *Life in the Cold,* p. 88.
6. R. K. Maxwell, J. Thorkelson, L. L. Rogers, and R. B. Brander, "The Field Energetics of Winter-Dormant Black Bear (*Ursus americanus*) in Northeastern Minnesota," *Canadian Journal of Zoology* 66 (1988): 2095–2103, and L. Rogers, "A Bear in Its Lair," *Natural History* 90 (1981): 64–70.
7. M. Aleksiuk, "Manitoba's Fantastic Snake Pits," *National Geographic* vol. 148, no. 5 (1975): 715–23.

Chapter 2. The Changing Snowpack

1. P. J. Marchand, "Light Extinction under a Changing Snowcover," in J. F. Merritt, ed., *Winter Ecology of Small Mammals.* Carnegie Museum of Natural History, Spec. Pub. 10 (Pittsburgh, 1984), 33–37.
2. C. W. Thomas, "On the Transfer of Visible Radiation through Sea Ice and Snow," *Journal of Glaciology* 4 (1963): 481–84; and H. Curl, J. T. Hardy, and R. Ellermeier, "Special Absorption of Solar Radiation in Alpine Snowfields," *Ecology* 53(1972): 1189–94.
3. Curl et al., "Spectral Absorption," 1189–94.
4. S. G. Richardson, and F. B. Salisbury, "Plant Responses to the Light Penetrating Snow," *Ecology* 58(1977): 1152–58.

5. It should be noted in passing that one frequently cited report in the literature of small mammal ecology contains conflicting evidence: L. N. Evenden, and W. A. Fuller, "Light Alteration Caused by Snow and Its Importance to Subnivean Rodents," *Canadian Journal of Zoology* 50 (1972): 1023–32. To measure light quality under snow; Evernden and Fuller employed a color temperature device in which the sensor temperature varied as a function of the wavelength of light absorbed. Using this technique, they observed an apparent shift in light penetration toward the red end of the spectrum under snow of increasing depth. The authors also studied the response of captive voles in a laboratory to light of different color and found that red light inhibited sexual maturation whereas blue or white light stimulated it. Putting the two results together suggested a mechanism for regulating reproductive activity under the snow. Unfortunately, the color temperature device appears to have been responding to something other than light (probably the snowpack temperature gradient). Furthermore, the authors indicate that the voles under the red light regime in the laboratory were possibly responding to a perceived total darkness rather than to color per se, since they appeared blind to the researcher's hand.

Chapter 3. Plants and the Winter Environment

1. J. R. Havis, "Water Movement in Woody Stems during Freezing, *Cryobiology* 8 (1971): 581–85.
2. P. L. Steponkus, and S. C. Wiest, "Plasma Membrane Alterations Following Cold-Acclimation and Freezing," in P. H. Li and A. Sakai, eds., *Plant Cold Hardiness and Freezing Stress: Mechanisms and Crop Implications* (New York: Academic Press, 1978).
3. J. P. Palta, and P. H. Li, "Cell Membrane Properties in Relation to Freezing Injury," in P. H. Li and A. Sakai, eds., *Plant Cold Hardiness and Freezing Stress: Mechanisms and Crop Implications* (New York: Academic Press, 1978).
4. P. J. Quinn, "A Lipid-Phase Separation Model of Low-Temperature Damage to Biological Membranes," *Cryobiology* 22 (1985): 128–46.
5. C. J. Weiser, "Cold Resistance and Injury in Woody Plants," *Science* (1970): 1269–78.
6. J. M. Schmitt, M. J. Schramm, H. Pfanz, S. Coughlan, and U. Heber, "Damage to Choloroplast Membranes during Dehydration and Freezing," *Cryobiology* 22 (1985): 93–104.

7. A. G. Hirsh, R. J. Williams, and H. T. Meryman, "A Novel Method of Natural Cryoprotection," *Plant Physiology* 79 (1985): 41–56.
8. W. C. White and C. J. Weiser, "The Relation of Tissue Desiccation, Extreme Cold, and Rapid Temperature Fluctuations to Winter Injury of American Arborvitae, *"American Society of Horticultural Science* 85 (1964): 554–63.
9. P. Wardle, "Winter Desiccation of Conifer Needles Simulated by Artificial Freezing," *Arctic and Alpine Research* 13 (1981): 419–23.
10. V. C. LaMarche and H. A. Mooney, "Recent Climatic Change and Development of the Bristlecone Pine (*Pinus longaeva* Bailey) Krummholz Zone, Mt. Washington, Nevada," *Arctic and Alpine Research* 4 (1972): 61–72.
11. A. Kacperska-Palacz, "Mechanism of Cold Acclimation in Herbaceous Plants," in P. H. Li and A. Sakai, eds., *Plant Cold Hardiness and Freezing Stress: Mechanisms and Crop Implications* (New York: Academic Press, 1978).
12. Weiser, "Cold Resistance," 1269–78.
13. Ibid.
14. F. B. Salisbury and C. W. Ross, *Plant Physiology* (Belmont, Ca.: Wadsworth Publishing Co., 1985).
15. Weiser, "Cold Resistance," 1269–78.
16. U. Heber, "Freezing Injury in Relation to Loss of Enzyme Activities and Protection against Freezing," *Cryobiology* 5 (1968): 188–201.
17. M. Senser, "Frost Resistance in Spruce (*Picea abies* (L.) Karst): III. Seasonal Changes in the Phospho- and Galacto-Lipids of Spruce Needles," *Zeitschrift für Pflanzenphysiologie* 105 (1982): 229–39.
18. Ibid.
19. Kacperska-Palacz, "Mechanism of Cold Acclimation."
20. W. Larcher and H. Bauer, "Ecological Significance of Resistance to Low Temperature," in O. L. Lange, P. S. Nobel, C. B. Osmond, and H. Ziegler, eds., *Physiological Plant Ecology I.* Encyclopedia of Plant Physiology, New Series, vol. 12a (Berlin: Springer-Verlag, 1981), 403–37.
21. Weiser, "Cold Resistance," 1269–78.
22. Larcher and Bauer, "Ecological Significance," 403–37.
23. S. Eiga and A. Sakai, "Altitudinal Variation in Freezing Resistance of Saghalien Fir (*Abies sachalinensis*)," *Canadian Journal of Botany* 62 (1984): 156–60.
24. A. F. W. Schimper, *Plant-Geography upon a Physiological Basis,* trans. W. R. Risher (Oxford: Clarendon, 1903).

25. W. Tranquillini, *Physiological Ecology of the Alpine Timberline* (Berlin: Springer-Verlag, 1979).
26. A. Sakai, "Mechanism of Desiccation Damage of Conifers Wintering in Soil Frozen Areas," *Ecology* 51 (1970): 657–64.
27. P. S. Nobel, *Biophysical Plant Physiology and Ecology* (San Francisco: W. H. Freeman and Co., 1983).
28. P. Holmgren, P. G. Jarvis, and M. S. Jarvis, "Resistances to Carbon Dioxide and Water Vapor Transfer in Leaves of Different Plant Species," *Physiologia Plantarum* 18 (1965): 557–73.
29. Nobel, *Biophysical Plant Physiology.*
30. P. J. Marchand and B. F. Chabot, "Winter Water Relations of Treeline Plant Species on Mt. Washington, New Hampshire." *Arctic and Alpine Research* 10 (1978): 105–16.
31. Ibid., 105–16.
32. D. T. Kincaid and E. E. Lyons, "Winter Water Relations of Red Spruce on Mount Monadnock, New Hampshire," *Ecology* 62 (1981): 1155–61.
33. Marchand and Chabor, "Winter Water Relations," 105–16.
34. H. T. Hammel, "Freezing of Xylem Sap without Cavitation," *Plant Physiology* 42 (1967): 55–66.
35. S. C. Gregory and J. A. Petty, "Valve Action of Bordered Pits in Conifers," *Journal of Experimental Botany* 24 (1973): 763–67.
36. Havis, "Water Movement in Woody Stems," 581–85.
37. G. Hygen, "Water Stress in Conifers during Winter," in B. Slavik, ed., *Water Stress in Plants* (The Hague: W. Junk, 1963).
38. N. Polunin, "Conduction through Roots in Frozen Soil," *Nature* 132 (1933): 313–14.
39. P. W. Owston, J. L. Smith, and H. G. Halverson, "Seasonal Water Movement in Tree Stems," *Forest Science* 18 (1972): 266–72.
40. Nobel, *Biophysical Plant Physiology* 74 and 483.
41. Ibid., 514.
42. Marchand and Chabot, "Winter Water Relations," 105–16.
43. U. Benecke and W. M. Havranek, "Gas-Exchange of Trees at Altitudes Up to Timberline, Craigieburn Range, New Zealand," in U. Benecke and M. R. Davis, eds., *Mountain Environments and Subalpine Tree Growth,* New Zealand Forest Service Technical Paper No. 70 (1980), 195–212.
44. Tranquillini, *Physiological Ecology.*
45. E. D. Schulze, H. A. Mooney, and E. L. Dunn, "Wintertime Photosynthesis of Bristlecone Pine (*Pinus aristata*) in the White Mountains of California," *Ecology* 48 (1967): 1044–47.

46. J. F. Chabot and B. F. Chabot, "Developmental and Seasonal Patterns of Mesophyll Ultrastructure in *Abies balsamea*," *Canadian Journal of Botany* 53 (1975): 295–304.
47. Marchand and Chabot, "Winter Water Relations," 105–16.
48. L. L. Tieszen, "Photosynthetic Competence of the Subnivean Vegetation of an Arctic Tundra," *Arctic and Alpine Research* 6 (1974): 253–56.
49. E. Mäenpää, "Photosynthesis and Respiration of Bilberry (*Vaccinium myrtillus* L.) under Winter Conditions," unpublished manuscript.
50. Tranquillini, *Physiological Ecology.*
51. T. O. Perry, "Winter-Season Photosynthesis and Respiration by Twigs and Seedlings of Deciduous and Evergreen Trees," *Forest Science* 17 (1971): 41–43.
52. Ibid.
53. M. Schaedle and K. C. Foote, "Seasonal Changes in the Photosynthetic Capacity of *Populus tremuloides* Bark," *Forest Science* 17 (1971): 308–13.
54. S. Eurola and A. Huttunen, "Riisitunturi," *Oulanka Reports* 5 (1984): 65–67.
55. J. L. Hadley and W. K. Smith, "Wind Effects on Needles of Timberline Conifers: Seasonal Influence on Mortality," *Ecology* 67 (1986): 12–19.
56. P. J. Marchand, F. L. Goulet, and T. C. Harrington, "Death by Attrition: An Hypothesis for Wave-Mortality of Subalpine *Abies balsamea*," *Canadian Journal of Forest Research* 16 (1986): 591–96.

Chapter 4. Animals and the Winter Environment

1. D. D. Feist, "Metabolic and Thermogenic Adjustments in Acclimatization of Subarctic Alaskan Red-backed Voles," in J. F. Merritt, ed., *Winter Ecology of Small Mammals.* Carnegie Museum of natural History, Spec. Pub. 10 (Pittsburgh, 1984), 131–38.
2. J. O. Wolf and W. Z. Lidicker, Jr., "Communal Winter Nesting and Food Sharing in Taiga Voles," *Behavioral Ecology and Sociobiology* 9 (1981): 237–40.
3. D. M. Madison, "Group Nesting and Its Ecological and Evolutionary Significance in Overwintering Microtine Rodents," in J. F. Merritt, ed., *Winter Ecology of Small Mammals.* Carnegie Museum of Natural History, Spec. Pub. 10 (Pittsburgh, 1984), 267–74.
4. Madison, "Group Nesting," 267–74.
5. L. Irving. *Arctic Life of Birds and Mammals, Including Man* (Berlin: Springer-Verlag, 1972).

6. Ibid.
7. J. S. Hart and O. Heroux, "A Comparison of Some Seasonal and Temperature-Induced Changes in *Peromyscus:* Cold Resistance, Metabolism, and Pelage Insulation," *Canadian Journal of Zoology* 31 (1953): 528–34.
8. Irving, *Arctic Life.*
9. B. A. Wunder, "Strategies for, and Environmental Cueing Mechanisms of, Seasonal Changes in Thermoregulatory Parameters of Small Mammals," in J. F. Merritt, ed., *Winter Ecology of Small Mammals.* Carnegie Museum of Natural History, Spec. Pub. 10 (Pittsburgh, 1984), 165–72.
10. Irving, *Arctic Life.*
11. S. A. Richards, *Temperature Regulation* (London: Wykeham Publications, 1973).
12. Schmidt-Nielsen, *Animal Physiology,* 291.
13. Feist, "Metabolic and Thermogenic Adjustments," 131–38.
14. Wunder, "Strategies for, and Environmental Cueing Mechanisms of, Seasonal Changes" 165–72; and J. F. Merritt, "Winter Survival Adaptations of the Short-tailed Shrew (*Blarina brevicauda*) in an Appalachian Montane Forest," *Journal of Mammalogy* 67 (1986): 450–64.
15. Feist, "Metabolic and Thermogenic Adjustments," 131–38.
16. V. Nolan, Jr., and E. D. Ketterson, "An Analysis of Body Mass, Wing Length, and Visible Fat Deposits of Dark-eyed Juncos Wintering at Different Latitudes," *Wilson Bulletin* 95 (1983): 603–20.
17. C. Carey, W. R. Dawson, L. C. Maxwell, and J. A. Faulkner, "Seasonal Acclimatization to Temperature in Cardueline Finches. II. Changes in Body Composition and Mass in Relation to Season and Acute Cold Stress," *Journal of Comparative Physiology* 125 (1978): 101–13.
18. Nolan and Ketterson, "An Analysis of Body Mass," 603–20.
19. W. R. Dawson and C. Carey, "Seasonal Acclimatization to Temperature in Cardueline Finches. I. Insulative and Metabolic Adjustments," *Journal of Comparative Physiology* 112 (1976): 317–33.
20. G. C. West and J. S. Hart, "Metabolic Responses of Evening Grosbeaks to Constant and to Fluctuating Temperatures," *Physiological Zoology* 39 (1966): 171–84.
21. G. C. West, "Shivering and Heat Production in Wild Birds," *Physiological Zoology* 38 (1965): 111–20.
22. K. Korhonen, "Microclimate in the Snow Burrows of Willow Grouse (*Lagopus lagopus*)," *Ann. Zool. Fennici* 17 (1980): 5–9.

23. W. A. Buttemer, "Energy Relations of Winter Roost-Site Utilization by American Goldfinches (*Carduelis tristis*)," *Oecologia* 68 (1985): 126–32.
24. S. B. Chaplin, "The Physiology of Hypothermia in the Black-capped Chickadee, *Parus atricapillus*," *Journal of Comparative Physiology* 112 (1976): 335–44; and West, "Shivering and Heat Production," 111–20.
25. M. W. Schwan and D. D. Williams, "Temperature Regulation in the Common Raven of Interior Alaska," *Comparative Biochemical Physiology* 60a (1978): 31–36; and West, "Shivering and Heat Production," 111–20.
26. C. Carey and R. L. Marsh, "Shivering Finches," *Natural History* 90 (1981): 58–63; and Carey et al., "Seasonal Acclimatization."
27. Carey and Marsh, "Shivering Finches," 58–63.
28. Chaplin, "Physiology of Hypothermia," 335–44.
29. Wunder, "Strategies for, and Environmental Cueing Mechanisms of, Seasonal Changes," 165–72.
30. Feist, "Metabolic and Thermogenic Adjustments," 131–38.
31. R. W. Coles, "Thermoregulatory Function of the Beaver Tail," *American Zoologist* 9 (1969): 203a.
32. D. L. Kilgore, Jr., and K. Schmidt-Nielsen, "Heat Loss from Ducks" Feet Immersed in Cold Water," *Condor* 77 (1975): 475–78.
33. J. H. Brown and R. C. Lasiewski, "Metabolism of Weasels: The Cost of Being Long and Thin," *Ecology* 53 (1972): 939–43.
34. D. C. Ure, "Autumn Mass Dynamics of Red-backed Voles (*Clebrionomys gapperi*) in Colorado in Relation to Photoperiod Cues and Temperature," in J. F. Merritt, ed., *Winter Ecology of Small Mammals,* Carnegie Museum of Natural History, Spec. Pub. 10 (Pittsburgh: 1984), 193–200.
35. B. Wunder, personal communication.
36. J. Zejda, personal communication.
37. B. K. McNab, "On the Ecological Significance of Bergmann's Rule," *Ecology* 52 (1971): 845–54.
38. Ibid.
39. W. J. Hamilton, III, and F. Heppner, "Radiant Solar Energy and the Function of Black Homeotherm Pigmentation: An Hypothesis," *Science* 155 (1967): 196–97.
40. W. J. Hamilton, III, *Life's Color Code* (New York: McGraw-Hill Book Co., 1973), 29–83.

41. Irving, *Arctic Life.*
42. B. Heinrich, "Thermoregulation by Winter-Flying Endothermic Moths," *Journal of Experimental Biology* 127 (1987): 313–32.
43. G. M. Courtin, J. D. Shorthouse, and R. J. West, "Energy Relations of the Snow Scorpionfly *Boreus brumalis* (Mecoptera) on the Surface of the Snow," *Oikos* 43 (1984): 241–45.
44. R. E. Lee, Jr., cautions that the question of intracellular freezing in insects is in need of additional systematic examination with advanced cryobiologic techniques. See discussion and references in R. E. Lee, Jr., and D. L. Denlinger, eds., "Insects at Low Temperature" (New York: Chapman and Hall, 1991), 35.
45. K. B. Storey and J. M. Storey, "Freeze Tolerance and Freeze Avoidance in Ectotherms," in C. H. Wang, ed., *Comparative and Environmental Physiology 4: Animal Adaptation to Cold* (Berlin: Springer-Verlag, 1989), 51–82.
46. Ibid.
47. Ibid.
48. Ibid.
49. R. E. Lee, Jr., personal communication.
50. Numerous studies summarized by R. J. C. Cannon and W. Block, "Cold Tolerance of Microarthropods," *Biological Review* 63 (1988): 23–77.
51. L. G. Neven, J. G. Duman, J. M. Beals, and F. J. Castellino, "Overwintering Adaptations of the Stag Bettle, *Ceruchus piceus:* Removal of Ice Nucleators in the Winter to Promote Supercooling," *Journal of Comparative Physiology* 156 (1986): 707–16.
52. J. G. Baust and K. E. Zachariassen, "Seasonal Active Cell Matrix Associated Ice Nucleators in an Insect," *Cryo-Letters* 4 (1983): 65–71.
53. J. Rickards, J. J. Kelleher, and K. B. Storey, "Strategies of Freeze Avoidance in Larvae of the Goldenrod Gall Moth, *Epiblema scudderiana:* Winter Profiles of a Natural Population," *Journal of Insect Physiology* 33 (1987): 443–50.
54. R. A. Ring, "The Physiology and Biochemistry of Cold Tolerance in Arctic Insects," *Journal of Thermal Biology* 6 (1981): 219–29.
55. J. Duman and K. Horwath, "The Role of Hemolymph Proteins in the Cold Tolerance of Insects," *Annual Review of Physiology* 45 (1983): 261–70.
56. Storey and Storey, "Freeze Tolerance and Avoidance," 51–82.
57. Ring, "Physiology and Biochemistry," 219–29.
58. Ibid.

59. Freeze tolerance in the garter snake was reported by J. P. Costanzo, D. L. Claussen, and R. E. Lee, Jr., "Natural Freeze Tolerance in a Reptile," *Cryo-Letters* 9 (1988): 380–85. References to freeze tolerance in other ectotherms are found in Storey and Storey, "Freeze Tolerance and Avoidance."
60. J. G. Baust and R. E. Lee, Jr., "Divergent Mechanisms of Frost Hardiness in Two Populations of the Gall Fly, *Eurosta solidaginis,*" *Journal of Insect Physiology* 27 (1981): 485–90.
61. Storey and Storey, "Freeze Tolerance and Avoidance," 51–82.
62. J. G. Duman, J. P. Morris, and F. J. Castellino, "Purification and Composition of an Ice Nucleating Protein from Queens of the Hornet, *Vespula maculata,*" Journal of Comparative Physiology 154 (1984): 79–83.
63. R. E. Lee, Jr., and E. A. Lewis, "Effect of Temperature and Duration of Exposure on Tissue Ice Formation in the Gall Fly, *Eurosta solidaginis* (Diptera, Tephritidae)," *Cryo-Letters* 6 (1985): 25–34.
64. Storey and Storey, "Freeze Tolerance and Avoidance," 51–81.
65. Ring, "Physiology and Biochemistry," 219–29.
66. J. G. Duman, "Change in Overwintering Mechanism of the Cucujid Beetle, *Cucujus clavipes,*" *Journal of Insect Physiology* 30 (1984): 235–39.
67. Storey and Storey, "Freeze Tolerance and Avoidance," 51–81.
68. Ibid.
69. Ibid.
70. W. D. Schmid, "Survival of Frogs in Low Temperature," *Science* 215 (1982): 697–98.
71. J. R. Layne, Jr. and R. E. Lee, Jr., "Freeze Tolerance and the Dynamics of Ice Formation in Wood Frogs (*Rana sylvatica*) from Southern Ohio," *Canadian Journal of Zoology* 65 (1987): 2062–65.
72. K. B. Storey and J. M. Storey, "Triggering of Cryoprotectant Synthesis by the Initiation of Ice Nucleation in the Freeze Tolerant Frog, *Rana sylvatica,*" *Journal of Comparative Physiology* 156 (1985): 191–95.
73. J. R. Layne, Jr., R. E. Lee, Jr., and T. L. Heil, "Freezing-induced Changes in the Heart Rate of Wood Frogs (*Rana sylvatica*)," *American Journal of Physiology* 257 (1989): R1046–49.
74. Ibid.
75. Ibid.
76. R. E. Lee, Jr., personal communication.
77. Layne et al., "Freezing-induced Changes."
78. R. E. Lee, Jr., "Insect Cold Hardiness: To Freeze or Not to Freeze," *BioScience* 39 (1989): 308–13.

79. P. J. Quinn, "A Lipid-phase Separation Model of Low-temperature Damage to Biological Membranes," *Cryobiology* 22 (1985): 128–46.
80. Lee, "Insect Cold-hardiness," 308–13.
81. R. E. Lee, Jr., C.-P. Chen, and D. L. Denlinger, "A Rapid Cold-hardening Process in Insects," *Science* 238 (1987): 1415–17.
82. R. Ziegler and G. R. Wyatt, "Phosphorylase and Glycerol Production Activated by Cold in Diapausing Silkmoth Pupae," *Nature* 254 (1975): 622–23.

Chapter 5. Life under Ice

1. M. W. Oswood, L. K. Miller, and J. G. Irons III, "Overwintering of Freshwater Benthic Macroinvertebrates," in R. E. Lee, Jr. and D. L. Denlinger, eds., *Insects at Low Temperature* (New York: Chapman and Hall, 1991), 360–5.
2. Ibid.
3. J. T. Eastman and A. L. DeVries, "Antarctic Fishes," *Scientific American* 255 (1986): 106–14.
4. A. L. DeVries, "Antifreeze Peptides and Glycopeptides in Cold-water Fishes," *Annual Review of Physiology* 45 (1983): 245–60.
5. Ibid. See numerous references to the work of DeVries and his associates.
6. Eastman and DeVries, "Antarctic Fishes," 110.
7. This temperature response is often indicated by a "Q_{10}" (for 10° C change) coefficient. A metabolic process with a Q_{10} equal to 2.0, as in this example, would be considered normally temperature sensitive, based on chemical kinetics alone. A Q_{10} of less than 2 would indicate relative insensitivity to temperature, while a Q_{10} on the order of 3 to 4 would indicate unusual sensitivity, suggesting more than strict chemical kinetics controlling the reaction.
8. $Q_{10} = 4.1$ in this case (see previous note).
9. P. J. Walsh, G. D. Foster, and T. W. Moon, "The Effects of Temperature on Metabolism of the American Eel (*Anquilla rostratae* (Le Seur): Compensation in the Summer and Torpor in the Winter," *Physiological Zoology* 56 (1983): 532–40.
10. J. R. Hazel, "Cold-adaptation in Ectotherms: Regulation of Membrane Function and Cellular Metabolism," in C. H. Wang, ed., *Advances in Comparative and Environmental Physiology 4: Animal Adaptation to Cold* (Berlin: Springer-Verlag, 1989), 1—50.
11. Ibid. See numerous references in this work.

12. K. Schmidt-Nielson, *Animal Physiology.*
13. K. P. Rao, "Rate of Water Propulsion in *Mytilus californianus* as a Function of Latitude," *Biological Bulletin* 104 (1953): 171–81. Figure 4 in this article (the basis for figure 51 in this book) is redrawn from R. W. Bullard, "Animals in Aquatic Environments: Annelids and Molluscs," in C. G. Wilber, E. F. Adalph, and D. B. Dill, eds., *Handbook of Physiology, Section 4: Adaptation to the Environment* (Washington, D.C.: American Physiological Society, 1964), fig. 1, 684.
14. P. L. Jones and B. D. Sidell, "Metabolic Responses of Striped Bass (*Marone saxatilis*) to Temperature Acclimation. II. Alterations in Metabolic Carbon Sources and Distributions of Fiber Types in Locomotory Muscle," *Journal of Experimental Zoology* 219 (1982): 163–71.
15. K. Schmidt-Nielson, *Animal Physiology.*
16. R. E. Gatten, Jr., "Anaerobiosis in Amphibians and Reptiles," *American Zoologist* 25 (1985): 945–54.
17. K. Schmidt-Nielson, *Animal Physiology,* 184–86.
18. R. T. Wright, "Dynamics of a Phytoplankton Community in an Ice-covered Lake," *Limnology and Oceanography* 9 (1964): 163–78.
19. C. W. Boylen and R. B. Sheldon, "Submergent Macrophytes: Growth under Winter Ice Cover," *Science* 194 (1976): 841–42.
20. P. J. Marchand, "Oxygen Evolution by *Elodea canadensis* under Snow and Ice Cover: A Case for Winter Photosynthesis in Subnivean Vascular Plants," *Aquilo Ser. Botanica* 23 (1985): 57–61.

Chapter 6. Plant-Animal Interactions

1. K. Danell, K. Huss-Danell, and R. Bergstrom, "Interaction between Browsing Moose and Two Species of Birch in Sweden," *Ecology* 66 (1985): 1867–78.
2. J. P. Bryant, G. D. Wieland, T. Clausen, and P. Kuropat, "Interactions of Snowshoe Hare and Feltleaf Willow in Alaska," *Ecology* 66 (1985): 1564–73.
3. J. P. Bryant, F. S. Chapin III, and D. R. Klein, "Carbon/Nutrient Balance of Boreal Plants in Relation to Vertebrate Herbivory," *Oikos* 40 (1983): 357–68.
4. P. B. Reichardt, J. P. Bryant, T. P. Clausen, and G. D. Wieland, "Defense of Winter-Dormant Alaska Paper Birch against Snowshoe hares," *Oecologia* 65 (1984): 58–69.
5. Ibid.
6. Bryant et al., "Interactions of Snowshoe Hare," 1564–73.

7. J. Tahvaninen, E. Helle, R. Julkunen-Titto, and A. Lavola, "Phenolic Compounds of Willow Bark as Deterrents against Feeding by Mountain Hare," *Oecologia* 65 (1985): 319–23.
8. Ibid.
9. L. B. Keith, "Role of Food in Hare Populations Cycles," *Oikos* 40 (1983): 385–95; and Bryant et al., "Carbon/Nutrient Balance," 357–68.
10. A. R. E. Sinclair, C. J. Krebs, J. N. M. Smith, and S. Boutin, "Population Biology of Snowshoe Hares. III. Nutrition, Plant Secondary Compounds and Food Limitation," *Journal of Animal Ecology* 57 (1986): 787–806.
11. A. R. E. Sinclair and J. N. M. Smith, "Do Plant Secondary Compounds Determine Feeding Preferences in Snowshoe Hares?" *Oecologia* 61 (1984): 403–10.
12. J. P. Bryant, J. Tahvanainen, M. Sulkinuja, R. Rulkunen-Tiitto, P. Reichardt, and T. Green, "Biogeographic Evidence for the Evolution of Chemical Defense by Boreal Birch and Willow against Mammalian Browsing," *American Naturalist* 134 (1989): 20–34.
13. Ibid.
14. Sinclair, et al., "Population Biology of Snowshoe Hares," 787–806.
15. K. Laine, and H. Henttunen, "The Role of Plant Production in Microtine Cycles in Northern Fennoscandia," *Oikos* 40 (1983): 407–18.
16. S. Eurola, H. Kyllonen, and K. Laine, "Plant Production and Its Relation to Climatic Conditions and Small Rodent Density in Kilpisjarvi Region (69° 05′ N, 20° 40′ E), Finnish Lapland," in J. F. Merritt, ed., *Winter Ecology of Small Mammals,* Carnegie Museum of Natural History, Spec. Pub. 10 (Pittsburgh, 1984), 121–30.
17. Oksanen and Oksanen, Rep. Kevo Subarctic Research Station 17 (1981): 7–31.
18. Laine and Henttunen, "The Role of Plant Production," 407–18.
19. Ibid.
20. Keith, "Role of Food," 385–95.
21. P. Berger, N. C. Negus, E. H. Sanders, and P. D. Gardner, "Chemical Triggering of Reproduction in *Microtus montanus,*" *Science* 214 (1981): 69–70; and P. Olsen, "The Stimulating Effect of a Phytohormone, Gibberellic Acid, on Reproduction of *Mus musculus,*" *Australian Wildlife Research* 8 (1981): 321–35.
22. F. B. Salisbury, "Light Conditions and Plant Growth under Snow," in J. F. Merritt, ed., *Winter Ecology of Small Mammals,* Carnegie Museum of Natural History, Spec. Pub. 10 (Pittsburgh: 1984), 39–50.

23. J. Tast and A. Kaikusale, "Winter Breeding of the Root Vole, *Microtus oeconomus,* in 1972/1973 at Kilpisjarvi, Finnish Lapland," *Am. Zool. Fennici* 13 (1976): 174–76.
24. W. A. Reiners, "Co_2 Evolution from the Floor of Three Minnesota Forests," *Ecology* 49 (1968): 471–83.
25. P. Havas and E. Mäenpää, "Evolution of Carbon Dioxide at the Floor of a *Hylocomium-Myrtillus* Type Spruce Forest," *Aquilo Ser. Botanica* 11 (1972): 4–22.
26. A. N. Formozov, *Snow Cover as an Integral Factor of the Environment and its Importance in the Ecology of Mammals and Birds,* trans. W. Prychodko and W. O. Pruitt, Jr., Boreal Institute, Occasional Pub. No. 1 (Edmonton: Univ. of Alberta, 1946).
27. Evernden and Fuller, "Light Alteration Caused by Snow," 1023–32.
28. K. Korhonen, "Ventilation in the Subnivean Tunnels of the Voles *Microtus agrestis* and *M. oeconomus,: Ann. Zool. Fennici* 17 (1980): 1–4.
29. See, for example, J. J. Kelley, Jr., D. F. Weaver, and B. P. Smith, "The Variation of Carbon Dioxide under the Snow in the Arctic," *Ecology* 49 (1968): 358–61 and C. E. Penny and W. O. Pruitt, Jr., "Subnivean Accumulation of CO_2 and Its Effects on Winter Distribution of Small Mammals," in J. F. Merritt, ed., *Winter Ecology of Small Mammals,* Carnegie Museum of Natural History, Spec. Pub. 10 (Pittsburgh: 1984), 373–80.
30. N. V. Basenina, "Influence of the Quality of Subnivean Air on the Arrangement of Winter Nests of Voles," *Zoologicheskii Zhurnal* 35 (1956): 940–42.
31. K. Aaltonen, S. Pasanen, and H. Aaloton, "Measuring System for CO_2 Concentration under Snow," *Aguilo Ser. Bot.* 23 (1985): 65–68.
32. Penny and Pruitt, "Subnivean Accumulation of CO_2," 373–80.
33. M. C. Young, P. J. Marchand, and C. A. Bryant, "Carbon Dioxide Accumulation beneath Snow and the Response of Voles to Elevated CO_2 under Simulated Subnivean Conditions," 1990, unpublished.

Chapter 7. Winter Profiles

1. R. J. Hudson and R. J. Christopherson, "Maintenance Metabolism," in R. J. Hudson and R. G. White, eds., *Bioenergetics of Wild Herbivores* (Boca Raton, FL: CRC Press, 1985), Ch. 6.
2. C. C. Gates and R. J. Hudson, "Effects of Posture and Activity on Metabolic Responses of Wapiti to Cold," *Journal of Wildlife Management* 43 (1979): 564–67.
3. Hudson and Christopherson, "Maintenance Metabolism," Ch. 6.

4. See discussion of the energetic consequences of size in A. R. French, "The Patterns of Mammalian Hibernation," *American Scientist* 76 (1988): 569–75.
5. A good summary of the literature is provided by G. Cederlund, R. Bergström, and F. Sandegren, "Winter Activity Patterns of Females in Two Moose Populations," *Canadian Journal of Zoology* 67 (1989): 1516–22.
6. R. J. Mackie, K. L. Hamlin, and D. F. Pac, "Mule Deer," in J. A. Chapman and G. A. Feldhamer, eds., *Wild Mammals of North America* (Baltimore: Johns Hopkins University Press, 1982), 866.
7. Ibid., 866.
8. H. M. Armleder, M. J. Waterhouse, D. G. Keisker, and R. J. Dawson, "Winter Habitat Use by Mule Deer in the Central Interior of British Columbia," *Canadian Journal of Zoology* 72 (1994): 1721–25.
9. Mackie et al., "Mule Deer," and A. N. Moen, "Energy Conservation by White-Tailed Deer in the Winter," *Ecology* 57 (1976): 192–98.
10. Moen, "Energy Conservation," 192–98. Calculations of food requirement assume a value of 8.4 kJ (2 kcal) per dry gram of browse material and a field water content of 50%.
11. An excellent summary is provided by S. G. Fancy and R. G. White, "Incremental Cost of Activity," in R. J. Hudson and R. G. White, eds., *Bioenergetics of Wild Herbivores* (Boca Raton, FL: CRC Press, 1985), Ch. 7.
12. Ibid.
13. M. A. Kautz, G. M. Van Dyne, L. H. Carpenter, and W. M. Mautz, "Energy Cost for Activities of Mule Deer Fawns," *Journal of Wildlife Management* 46 (1982): 704–10.
14. D. T. Brown and G. J. Doucet, "Temporal Changes in Winter Diet Selection by White-Tailed Deer in a Northern Deer Yard," *Journal of Wildlife Management* 55 (1991): 361–76.
15. P. B. Gray and F. A. Servello, "Energy Intake Relationships for White-Tailed Deer on Winter Browse Diets," *Journal of Wildlife Management* 59 (1995): 147–52.
16. Mackie et al., "Mule Deer," 862.
17. Considerable discussion on heat production and loss is given in K. L. Parker and C. T. Robbins, "Thermoregulation in Ungulates," in Hudson and White, *Bioenergetics of Wild Herbivores,* Ch. 8. Though effective peripheral cooling is suggestive of a countercurrent heat exchange mechanism (see pp. 115–17), the well-defined rete mirabile

found in the legs and tail of some animals has not been reported in deer. For additional information on the energy balance of deer in winter, the reader should consult W. M. Mautz, J. Kanter, and P. J. Pekins, "Seasonal Metabolic Rhythms of Captive Female White-Tailed Deer: A Reexamination," *Journal of Wildlife Management* 56 (1992): 656–61, and K. A. Worden and P. J. Pekins, "Seasonal Change in Feed Intake, Body Composition, and Metabolic Rate of White-Tailed Deer," *Canadian Journal of Zoology* 73 (1995): 452–57.

18. A. N. Moen and C. W. Severinghaus, "Hair Depths of the Winter Coat of White-Tailed Deer," *Journal of Mammalogy* 65 (1984): 497–99.
19. Ibid.
20. Parker and Robbins, "Thermoregulation in Ungulates," 161–82.
21. Mackie et al., "Mule Deer," 862.
22. P. Trayhurn, "Species Distribution of Brown Adipose Tissue: Characterization of Adipose Tissues from Uncoupling Protein and its mRNA," in C. Carey, G. L. Florant, B. A. Wunder, and B. Horwitz, eds., *Life in the Cold: Ecological, Physiological, and Molecular Mechanisms* (Boulder, CO: Westview Press, 1993), 361–68.
23. E. Armstrong, D. Euler, and G. Racey, "Winter Bed-Site Selection by White-Tailed Deer in Central Ontario," *Journal of Wildlife Management* 47 (1983): 880–83.
24. Parker and Robbins, "Thermoregulation in Ungulates," 161–82.
25. Mackie et al., "Mule Deer," 862.
26. L. D. Mech, R. E. McRoberts, R. O. Peterson, and R. E. Page, "Relationship of Deer and Moose Populations to Previous Winters' Snow," *Journal of Animal Ecology* 56 (1987): 615–27.
27. J. A. Harper, J. H. Harn, W. W. Bentley, and C. F. Yocum, "The Status and Ecology of the Roosevelt Elk in California," *Wildlife Monographs* 16 (Washington, DC: The Wildlife Society, 1967).
28. V. Geist, "Adaptive Behavioral Strategies," in J. W. Thomas and D. E. Toweill, eds., *Elk of North America: Ecology and Management* (Harrisburg, PA: Stackpole Books, 1982), 219–27.
29. J. M. Peek, "Elk," in Chapman and Feldhamer, *Wild Mammals of North America,* 851.
30. Ibid.
31. K. L. Parker, C. T. Robbins, and T. A. Hanley, "Energy Expenditure for Locomotion by Mule Deer and Elk," *Journal of Wildlife Management* 48 (1984): 474–88.

32. Ibid.
33. E. S. Telfer and J. P. Kelsall, "Adaptation of some Large North American Mammals for Survival in Snow," *Ecology* 65 (1984): 1828–34.
34. J. M. Sweeney and J. R. Sweeney, "Snow Depths Influencing Winter Movements of Elk," *Journal of Mammalogy* 65 (1984): 524–26.
35. L. A. Renecker and R. J. Hudson, "Seasonal Energy Expenditures and Thermoregulatory Responses of Moose," *Canadial Journal of Zoology* 64 (1986): 322–27.
36. J. W. Coady, "Moose," in Chapman and G. A. Feldhamer, *Wild Mammals of North America,* 902.
37. L. Irving, *Arctic Life of Birds and Mammals, Including Man* (Berlin: Springer-Verlag, 1972) and Renecker and Hudson, "Seasonal Energy Expenditures," 322–27.
38. Renecker and Hudson, "Seasonal Energy Expenditures," 322–27.
39. Coady, "Moose," 902.
40. Ibid.
41. Ibid.
42. M. P. Gillingham and D. R. Klein, "Late-Winter Activity Patterns of Moose (*Alces alces gigas*) in Western Alaska," *Canadian Journal of Zoology* 70 (1992): 293–99.
43. Coady, "Moose," 902.
44. K. L. Risenhoover, "Winter Activity Patterns of Moose in Interior Alaska," *Journal of Wildlife Management* 50 (1986): 727–34.
45. Ibid.
46. Renecker and Hudson, "Seasonal Energy Expenditures," 322–27, and W. L. Regelin, C. C. Schwartz, and A. W. Dranzmann, "Seasonal Energy Metabolism of Adult Moose," *Journal of Wildlife Management* 49 (1985): 388–93.
47. C. C. Schwartz, M. E. Hubbert, and A. W. Franzmann, "Energy Requirements of Adult Moose for Winter Maintenance," *Journal of Wildlife Management* 52 (1988): 26–33.
48. O. Hjeljord, B-E. Sæther, and R. Andersen, "Estimating Energy Intake of Free-Ranging Moose Cows and Calves Through Collection of Feces," *Canadian Journal of Zoology* 72 (1994): 1409–15.
49. G. Worthy, J. Rose, and F. Stormshak, "Anatomy and Physiology of Fur Growth: The Pelage Priming Process," in M. Novak, J. A. Baker, M. E. Obbard, and B. Malloch, eds., *Wild Furbearer Management and Conservation in North America* (Toronto: Ministry of Natural Resources, 1987), 827.

---

50. Irving, *"Arctic Life."*
51. L. C. Cuyler and N. A. Øritsland, "Metabolic Strategies for Winter Survival by Svalbard Reindeer," *Canadian Journal of Zoology* 71 (1993): 1787–92.
52. R. D. Boertje, "An Energy Model for Adult Female Caribou of the Denali Herd, Alaska," *Journal of Range Management* 38 (1985): 468–73.
53. E. H. McEwan and P. E. Whitehead, "Seasonal Changes in the Energy and Nitrogen Intake in Reindeer and Caribou," *Canadian Journal of Zoology* 48 (1970): 905–13.
54. F. L. Miller, "Caribou," in Chapman and Feldhamer, *Wild Mammals of North America,* 923.
55. S. G. Fancy and R. G. White, "Energy Expenditures by Caribou While Cratering in the Snow," *Journal of Wildlife Management* 49 (1985): 987–93.
56. H. Thing, "Behavior, Mechanics, and Energetics Associated With Winter Cratering by Caribou in Northwestern Alaska," *Biology Papers of the University of Alaska,* no. 18 (1977).
57. D. C. Thomas, R. H. Russell, E. Broughton, E. Edmonds, and A. Gunn, "Further Studies of Two Populations of Peary Caribou in the Canadian Arctic," *Canadian Wildlife Service Progress Note,* no. 80 (1977).
58. O. Murie, *A Field Guide to Animal Tracks* (Boston: Houghton Mifflin, 1975).
59. Fancy and White, "Energy Expenditures by Caribou," 987–93.
60. Miller, "Caribou," 923.
61. Cuyler and Øritsland, "Metabolic Strategies," 1787–92.
62. E. Reimers, T. Ringberg, and R. Sørumgård, "Body Composition of Svalbard Reindeer," *Canadian Journal of Zoology* 60 (1982): 812–21.
63. Boertje, "An Energy Model for Adult Female Caribou," 468–73.
64. Miller, "Caribou," 923.
65. P. L. Errington, *Muskrat Populations* (Ames: Iowa State University Press, 1963).
66. R. A. MacArthur and M. Aleksiuk, "Seasonal Microenvironments of the Muskrat (*Ondatra zibethicus*) in a Northern Marsh," *Journal of Mammalogy* 60 (1979): 146–54.
67. R. A. MacArthur, "Aquatic Mammals in Cold," in L. C. H. Wang, ed., *Advances in Comparative and Environmental Physiology: Animal Adaptation to Cold* (Berlin: Springer-Verlag, 1989), 289–325.
68. Ibid.

69. R. A. MacArthur, "Seasonal Patterns of Body Temperature and Activity in Free-Ranging Muskrats (*Ondatra zibethicus*)," *Canadian Journal of Zoology* 57 (1979): 25–33.
70. K. Johansen, "Buoyancy and Insulation in the Muskrat," *Journal of Mammalogy* 43 (1962): 64–8.
71. E. S. Fairbanks and D. L. Kilgore, Jr., "Post-Dive Oxygen Consumption of Restrained and Unrestrained Muskrats (*Ondatra zibethicus*)," *Comparative Biochemical Physiology* A59 (1978): 113–17. See also discussion in R. A. MacArthur, "Aquatic Thermoregulation in the Muskrat (*Ondatra zibethicus*): Energy Demands of Swimming and Diving," *Canadian Journal of Zoology* 62 (1984): 241–48.
72. H. J. Harlow, "The Influence of Hardarian Gland Removal and Fur Lipid Removal on Heat Loss and Water Flux to and from the Skin of Muskrats, *Ondatra zibethicus,*" *Physiological Zoology* 57 (1984): 349–56.
73. MacArthur, "Seasonal Patterns," 25–33, and R. A. MacArthur, "Gas Bubble Release by Muskrats Diving Under Ice: Lost Gas or a Potential Oxygen Pool?," *Journal of Zoology (London)* 226 (1992): 151–64.
74. MacArthur, "Aquatic Thermoregulation," 241–48.
75. Ibid.
76. R. A. MacArthur, "Seasonal Changes in the Oxygen Storage Capacity and Aerobic Dive Limits of the Muskrat (*Ondatra zibethicus*)," *Journal of Comparative Physiology B* 160 (1990): 593–99.
77. R. A. MacArthur, "Foraging Range and Aerobic Endurance of Muskrats Diving Under Ice," *Journal of Mammalogy* 73 (1992): 565–69.
78. MacArthur, "Seasonal Changes in Oxygen Storage Capacity," 593–99.
79. MacArthur, "Gas Bubble Release," 151–64.
80. Ibid.
81. Errington, *Muskrat Populations.*
82. Ibid.
83. F. Messier and J. A. Virgil, "Differential use of Bank Burrows and Lodges by Muskrats, *Ondatra zibethicus,* in a Northern Marsh Environment," *Canadian Journal of Zoology* 70 (1992): 1180–84.
84. N. S. Novakowski, "The Winter Bioenergetics of a Beaver Population in Northern latitudes," *Canadian Journal of Zoology* 45 (1967): 1107–18.
85. B. G. Slough, "Beaver Food Cache Structure and Utilization," *Journal of Wildlife Management* 42 (1978): 644–46.
86. Novakowski, "Winter Bioenergetics," 1107–18.

87. Ibid.
88. A. P. Dyck and R. A. MacArthur, "Daily Energy Requirements of Beaver (*Castor canadensis*) in a Simulated Winter Microhabitat," *Canadian Journal of Zoology* 71 (1993): 2131–35.
89. Novakowski, "Winter Bioenergetics," 1107–18.
90. M. Aleksiuk and I. M. Cowan, "Aspects of Seasonal Energy Expenditure in the Beaver (*Castor canadensis* Kuhl) at the Northern Limit of Its Distribution, *Canadian Journal of Zoology* 47 (1969): 471–81.
91. R. A. MacArthur and A. P. Dyck, "The Search for a Metabolic Depression in the Beaver: 'Lost Opportunities' or Red Herrings? A Reply to J. Bovet," *Canadian Journal of Zoology* 72 (1994): 569–71.
92. D. W. Smith, R. O. Peterson, T. D. Drummer, and D. S. Sheputis, "Over-Winter Activity and Body Temperature Patterns in Northern Beavers," *Canadian Journal of Zoology* 69 (1991): 2178–82.
93. A. P. Dyck and R. A. MacArthur, "Seasonal Patterns of Body Temperature and Activity in Free-Ranging Beaver (*Castor canadensis*)," *Canadian Journal of Zoology* 70 (1992): 1668–72.
94. R. A. MacArthur and A. P. Dyck, "Aquatic Thermoregulation of Captive and Free-Ranging Beavers (*Castor canadensis*)," *Canadian Journal of Zoology* 68 (1990): 2409–16.
95. J. Bovet and E. F. Oertli, "Free-Running Circadian Activity Rhythms in Free-Living Beaver (*Castor canadensis*)," *Journal of Comparative Physiology* 92 (1974): 1–10; C. L. Potvin and J. Bovet, "Annual Cycle of Patterns of Activity Rhythms in Beaver Colonies (*Castor canadensis*)," *Journal of Comparative Physiology* 98 (1975): 243–56; R. A. Lancia, W. E. Dodge, and J. S. Larson, "Winter Activity Patterns of Two Radio-Marked Beaver Colonies," *Journal of Mammalogy* 63 (1982): 598–606.
96. Referring to studies cited above by Smith et al., "Over-Winter Activity," Dyck and MacArthur, "Seasonal Patterns," Potvin and Bovet, "Annual Cycle of Patterns," and Lancia et al., "Winter Activity Patterns."
97. MacArthur and Dyck, "Aquatic Thermoregulation," 2409–16.
98. Dyck and MacArthur, "Seasonal Patterns," 1668–72.
99. MacArthur, "Gas Bubble Release," 151–64.
100. D. G. Reid, T. E. Code, A. C. H. Reid, and S. M. Herrero, "Spacing, Movements, and Habitat Selection of the River Otter in Boreal Alberta," *Canadian Journal of Zoology* 72 (1994): 1314–24. See also W. E. Melquist and Ana E. Dronkert, "River Otter," in M. Novak, J. A. Baker, M. E. Obbard, and B. Malloch, eds., *Wild Furbearer*

Management and Conservation in North America (Toronto: Ministry of Natural Resources, 1987), 626–41, for discussion.

101. W. E. Melquist and M. G. Hornocker, "Ecology of River Otters in West Central Idaho," *Wildlife Monographs* 83 (1983): 1–60.
102. Reid et al., "Spacing, Movements, and Habitat Selection," 1314–24.
103. Melquist and Hornocker, "Ecology of River Otters," 1–60.
104. F. J. Tarasoff, "Anatomical Adaptations in the River Otter, Sea Otter and Harp Seal with Reference to Thermal Regulation," in R. J. Harrison, ed., *Functional Anatomy of Marine Mammals,* Vol. 2 (New York: Academic Press, 1974), 111–41.
105. Ibid., 140.
106. Ibid., 133–34.
107. Ibid., 135.
108. Ibid.
109. See discussion in D. D. Feist and R. G. White, "Terrestrial Mammals in Cold," in L. C. H. Wang, ed., *Advances in Comparative and Environmental Physiology: Animal Adaptation to Cold* (Berlin: Springer-Verlag, 1989), 327–60.
110. R. T. Eberhardt, "Some Aspects of Mink-Waterfowl Relationships of Prairie Wetlands," *Prairie Naturalist* 5 (1973): 17–19.
111. Errington, *Muskrat Populations.*
112. N. Dunstone and R. J. O'Connor, "Optimal Foraging in an Amphibious Mammal. I. The Aqualung Effect," *Animal Behavior* 27 (1979): 1182–1194, and T. B. Poole and N. Dunstone, "Underwater Predatory Behavior of the American Mink (*Mustela vison*)." *Journal of Zoology (London)* 178 (1976): 395–412.
113. T. M. Williams, "Thermoregulation of the North American Mink During Rest and Activity in the Aquatic Environment," *Physiological Zoology* 59 (1986): 293–305.
114. Ibid.
115. S. W. Buskirk, H. J. Harlow, and S. C. Forrest, "Temperature Regulation in American Marten (*Martes americana*) in Winter," *National Geographic Research* 4 (1988): 208–18.
116. Ibid.
117. R. L. Marsh and W. R. Dawson, "Avian Adjustments to Cold," in L. C. H. Wang, ed., *Advances in Comparative and Environmental Physiology: Animal Adaptation to Cold* (Berlin: Springer-Verlag, 1989), 205–53.
118. H. Rintamäki, S. Saarela, A. Marjakangas, and R. Hissa, "Summer and Winter Temperature Regulation in the Black Grouse *Lyrurus tetrix,*" *Physiological Zoology* 56 (1983): 152–59.

119. R. Moss, "Gut Size, Body Weight, and Digestion of Winter Foods by Grouse and Ptarmigan," *Condor* 85 (1983): 185–93.
120. R. H. McBee and G. C. West, "Cecal Fermentation in the Willow Ptarmigan," *Condor* 71 (1969): 54–58.
121. Ibid.
122. J. K. Terres, *The Audubon Society Encyclopedia of North American Birds* (New York: Alfred A. Knopf, 1980), 449–55.
123. D. M. Keppie, "Snow Cover and the Use of Trees by Spruce Grouse in Autumn," *Condor* 79 (1977): 382–84.
124. L. N. Ellison, "Seasonal Social Organization and Movements of Spruce Grouse," *Condor* 75 (1973): 375–85.
125. A. V. Andreev, "Winter Adaptations in the Willow Ptarmigan," *Arctic* 44 (1991): 106–14.
126. V. G. Thomas, "Winter Diet and Intestinal Proportions of Rock and Willow Ptarmigan and Sharp-Tailed Grouse in Ontario," *Canadian Journal of Zoology* 62 (1984): 2258–63.
127. B. A. Pendergast and D. A. Boag, "Seasonal Changes in the Internal Anatomy of Spruce Grouse in Alberta," *The Auk* 90 (1973): 307–17, and R. Moss, "Gut Size."
128. L. Fenna and D. A. Boag, "Adaptive Significance of the Caeca in Japanese Quail and Spruce Grouse (Galliformes)," *Canadian Journal of Zoology* 52 (1974): 1577–84.
129. A. V. Andreev, "Ecological Energetics of Palaearctic Tetraonidae in Relation to Chemical Composition and Digestibility of Their Winter Diets," *Canadian Journal of Zoology* 66 (1988): 1382–88.
130. F. B. Gill, *Ornithology,* 2nd edition (New York: W. H. Freeman, 1995), 68.
131. V. G. Thomas, H. G. Lumsden, and D. H. Price, "Aspects of the Winter Metabolism of Ruffed Grouse (*Bonasa umbellus*) with Special Reference to Energy Reserves," *Canadian Journal of Zoology* 53 (1975): 434–40.
132. E. O. Höhn, "The 'Snowshoe Effect' of the Feathering on Ptarmigan Feet," *Condor* 79 (1977): 380–82.
133. Andreev, "Ecological Energetics," 1382–88.
134. A. Marjakangas, H. Rintamäki, and R. Hissa, "Thermal Responses in the Capercaillie *Tetrao urogallus* and the Black Grouse *Lyrurus tetrix* Roosting in the Snow," *Physiological Zoology* 57 (1984): 99–104.
135. Ibid.
136. Rintamäki et al., "Summer and Winter Temperature Regulation," 152–59.

---

137. B. S. Cade and R. W. Hoffman, "Differential Migration of Blue Grouse in Colorado," *The Auk* 110 (1993): 70–77.
138. P. J. Pekins, J. A. Gessaman, and F. G. Lindzey, "Field Metabolic Rate of Blue Grouse During Winter," *Canadian Journal of Zoology* 72 (1994): 227–31.
139. Cade and Hoffman, "Differential Migration," 70–77, and Pekins et al., "Field Metabolic Rate," 227–31.
140. J. E. Hines, "Social Organization, Movements, and Home Ranges of Blue Grouse in Fall and Winter," *Wilson Bulletin* 98 (1986): 419–32, and Pekins et al., "Field Metabolic Rate," 227–31.
141. P. J. Pekins, J. A. Gessaman, and F. G. Lindzey, "Winter Energy Requirements of Blue Grouse," *Canadian Journal of Zoology* 70 (1992): 22–24, and Pekins et al., "Field Metabolic Rate," 227–31.
142. V. G. Thomas, H. G. Lumsden, and D. H. Price, "Aspects of the Winter Metabolism of Ruffed Grouse (*Bonasa umbellus*) with Special Reference to Energy Reserves," *Canadian Journal of Zoology* 53 (1975): 434–40, and Pekins et al., "Winter Energy Requirements," 22–24.
143. G. C. West, "Bioenergetics of Captive Willow Ptarmigan Under Natural Conditions," *Ecology* 49 (1968): 1035–45; V. G. Thomas and R. Popko, "Fat and Protein Reserves of Wintering and Prebreeding Rock Ptarmigan from South Hudson Bay," *Canadian Journal of Zoology* 59 (1981): 1205–11; Thomas et al., "Aspects of Winter Metabolism," 434–40.
144. A comprehensive review of seasonal accumulation of energy reserves is provided in Marsh and Dawson, "Avian Adjustments to Cold," 205–53.

Chapter 8. Humans in Cold Places

1. H. C. Bazett, "The Regulation of Body Temperatures," in L. H. Newburgh, ed., *Physiology of Heat Regulation and the Science of Clothing* (Philadelphia: W. B. Saunders, 1949), 109–92. The clo is a unit devised to rate the insulative value of clothing; 1 clo will maintain a resting and sitting person with a metabolic rate of 209 joules/m^2/hr comfortable at 21° C with relative humidity less than 50% and no wind. See M. K. Yousef, "Effects of Climate Stresses on Thermoregulatory Processes in Man," *Experientia* 43 (1987): 14–19, for references and the computational formula for clo units.
2. J. LeBlanc, *Man in the Cold* (Springfield, Ill: C. C. Thomas, 1975).
3. J. LeBlanc, "Subcutaneous Fat and Skin Temperature," *Can. J. Biochem. Physiol.* 32 (1954): 354–59.

4. R. P. Clark and O. G. Edholm, *Man and His Thermal Environment* (London: Edward Arnold, 1985), 165.
5. A. J. F. Webster, "Adaptation to Cold," in D. Robertshaw, ed., *Physiology Series One, Volume 7, Environmental Physiology* (London: Butterworths, 1974).
6. K. E. Cooper, "The Role of Neural Inputs and Their Processing in Cold Adaptation," in L. S. Underwood, L. L. Tieszen, A. B. Callahan, and G. E. Folk, eds., *Comparative Mechanisms of Cold Adaptation* (New York: Springer-Verlag, 1991).
7. H. O. Garland, "Altered Temperature," in R. M. Case, ed., *Variations in Human Physiology* (Manchester, U.K.: Manchester University Press, 1985), 119.
8. S. A. Richards, *Temperature Regulation* (London: Wykeham Publications Ltd., 1973), 77.
9. Garland, "Altered Temperature," 119.
10. J. Werner, "Influences of Local and Global Temperature Stimuli on the Lewis-Reaction," *Pflygers Archiv* 367 (1977): 291–94.
11. T. Adams and R. E. Smith, "Effect of Chronic Local Cold Exposure on Finger Temperature Responses," *Journal of Applied Physiology* 17 (1962): 317.
12. See summary of literature by Webster, "Adaptation to Cold," 87.
13. Werner, "Influences of Local and Global Temperature," 292.
14. Ibid.
15. C. J. Eagan, "Introduction and Terminology: Habituation and Peripheral Tissue Adaptations," in E. F. Moran, *Human Adaptability* (Boulder, Colo.: Westview Press, 1982), 121.
16. L. K. Miller and L. Irving, "Local Reactions to Air Cooling in an Eskimo Population," *Journal of Applied Physiology* 17 (1962): 449–55.
17. J. LeBlanc, "Adaptation of Man to Cold," in L. C. H. Wang and J. W. Hudson, eds., *Strategies in Cold: Natural Torpidity and Thermogenesis* (New York: Academic Press, 1978), 708–11.
18. A. Holdcroft, *Body Temperature Control in Anaesthesea, Surgery and Intensive Care* (London: Bailliere Tindall, 1980).
19. O. G. Edholm and A. L. Bacharach, *The Physiology of Human Survival* (London: Academic Press, 1965).
20. See review by Webster, "Adaptation to Cold," 88.
21. LeBlanc, S. Dulac, J. Core, and B. Girard, "Autonomic Nervous System and Adaptation to Cold in Man," *Journal of Applied Physiology* 39 (1975): 181–86.
22. Cooper, "The Role of Neural Inputs," 75–79.

23. Ibid.
24. Holdcroft, *Body Temperature Control.*
25. Ibid.
26. M. K. Yousef, "Effects of Climatic Stresses on Thermoregulatory Processes in Man," *Experientia* 43 (1987): 14–19; Holdcroft, *Body Temperature Control,* and others.
27. Ibid.
28. H. T. Hammel, "Terrestrial Animals in Cold: Recent Studies of Primitive Man," in D. B. Dill, ed., *Adaptation to the Environment* (Washington, D.C.: American Physiological Society, 1964), 413–34.
29. Garland, "Altered Temperature," 116.
30. C. C. VanWie, "Physiological Responses to Cold Environments," in J. D. Ives and R. G. Barry, eds., *Arctic and Alpine Environments* (London: Methuen, 1974).
31. Clark and Edholm, "Man and His Thermal Environment," 57.
32. J. M. Heaton, "The Distribution of Brown Adipose Tissue in the Human," *Journal of Anatomy* 112 (1972): 35–39.
33. P. Huttunen, J. Hirronen, and V. Kinnula, "The Occurrence of Brown Adipose Tissue in Outdoor Workers," *European Journal of Applied Physiology* 46 (1981): 339–45.
34. Holdcroft, "Body Temperature Control," 10.
35. Ibid.
36. Richards, "Temperature Regulation."
37. See references cited in D. D. Feist and R. G. White, "Terrestrial Mammals in Cold," in L. C. H. Wang, ed., *Advances in Comparative and Environmental Physiology 4: Animal Adaption to Cold* (Berlin: Springer-Verlag, 1989), 353.
38. L. Kavaler, *Freezing Point* (New York: John Day Co., 1970).
39. L. Irving, *Arctic Life of Birds and Mammals, Including Man* (Berlin: Springer-Verlag, 1972).
40. D. D. Jackson, "Up in the 'Cold Lab' Human Guinea Pigs Shiver for Science," *Smithsonian* 19 (1986): 101–09.
41. H. Benson, J. W. Lehmann, M. S. Malhotra, R. F. Goldman, J. Hopkins, and M. D. Epstein, "Body Temperature Changes During the Practice of g Tum-mo Yoga," *Nature* 295 (1982): 234–36.
42. W. Selvamurthy, U. S. Ray, K. S. Hedge, and R. P. Sharma, "Physiological Responses to Cold (10°C) in Men After Six Months' Practice of Yoga Exercises," *Int. J. Biometeorol,* 32 (1988): 188–93.
43. Irving, *Arctic Life.*

GLOSSARY

Acclimation. Physiological adjustment to changing environmental conditions, especially to increasing or decreasing temperature. This term is used in the plant literature in reference to seasonal changes in cold tolerance, but animal researchers restrict its use to shorter-term adjustments, usually observed under laboratory conditions (*see* Acclimatization).

Acclimatization. Seasonal or long-term physiological adjustment, usually in response to temperature changes. Preferred in this context by animal researchers over the term "acclimation."

Ambient. Referring to near surroundings. Characteristic of the immediate environment in which a plant or animal operates.

Anchor ice. Ice that accumulates on underwater surfaces in streams, usually as a result of the agglomeration of frazil ice particles.

Arteriovenous anastomosis. A small channel connecting arterioles and venules in the extremities and serving as a bypass of downstream capillary beds during vasoconstriction.

Basal metabolic rate (BMR). The lowest rate of metabolism shown by a homeotherm while fasting and at rest within its thermoneutral zone.

Benthic. Pertaining to bottom waters or substrate.

Boundary layer. Thin shell of air surrounding an object and having properties (e.g., temperature, humidity) intermediate between the surface of the object and the bulk atmosphere around it.

Brown fat. Adipose tissue characterized by a yellow to light brown color, having a high concentration of mitochondria and hence capable of high oxidation rates and heat production.

Caecum. Long, tubular sac branching off from the lower intestine, serving as a fermentation chamber in nonruminant herbivores.

Cavitation. Formation of a gas-filled bubble or cavity in a liquid, often occurring in the narrow capillaries of a plant as a result of the dissolution of gases upon freezing.

Cervid. Any member of the deer family, Cervidae.

Circadian rhythm. Endogenous biological cycle with a recurrence interval of approximately 24 hours.

Conduction. Transfer of heat via molecular collisions.

Constructive metamorphism. The process in which ice crystals favorably situated in the snowpack grow by accretion of water onto their surfaces.

Contour feather. Any of the outer feathers of the head, neck, and body, and flight feathers of the wings and tail.

Convection. Transfer of heat via a moving fluid.

Cuticle. A layer of wax covering the outer surface of a plant leaf, serving primarily to reduce water loss.

Denaturation. Alteration or loss of normal biotic function.

Depth Hoar. Brittle ice crystals, often hollow and cuplike, formed in warmer layers of the snowpack as a result of continuous vapor loss from their surfaces.

Desiccation. Extreme and damaging water loss.

Destructive metamorphism. The process in which new-fallen snow crystals lose their delicate structure by a redistribution of internal energy, and coalesce into rounded ice grains.

Diffuse porous. A characteristic ring pattern in the wood (xylem tissue) of some broadleaf tree species in which individual conducting cells are of small size and vary little in diameter from early spring to late summer, thus showing only a relatively faint line of demarcation between one year's growth and the next (cf. Ring porous).

Ectotherm. An animal that derives heat from external sources, for example by basking in sunlight, rather than producing heat internally through oxidative metabolism (similar in meaning to poikilotherm).

Endotherm. An animal that is capable of generating a significant amount of heat through muscular activity or other metabolic processes. The term encompasses homeotherms as well as some poikilotherms (see respective definitions).

Exothermic. A reaction or change of state in which heat is given off.

Extracellular. Outside of the cell (synonomous with Intercellular).

Fat body. Tissue of indeterminate form distributed throughout the body of insects and serving as a nutrient reserve.

Frazil ice. Small disk-shaped platelets of ice suspended in water.

Free-running. Not maintaining internal biological rhythms on a 24-hour cycle.

Freeze concentrated. An increase in the concentration of dissolved substances as a result of their having been excluded from growing ice crystals.

Glycolosis. The breakdown of carbohydrate in the absence of oxygen to produce ATP and lactic acid.

Hemolymph. The circulating body fluid of an insect, equivalent to blood and lymph of higher life-forms.

Herbaceous. Nonwoody.

Hibernaculum. Any refuge used over winter by a hibernating animal.

Homeotherm. An animal that maintains a body temperature within a high and relatively narrow range independently of the temperature of its surroundings. Often referred to as a "warm-blooded" animal.

Hunting response. See "Lewis reaction."

Hydrolysis. The addition of hydrogen and hydroxyl (OH) ions from water to a molecule, with its subsequent splitting into two or more simpler molecules.

Hydrophobic. Having little or no affinity for water (e.g., water repellent).

Hypothalamus. Specifically, the floor and sides of the brain just behind the attachment of the cerebral hemispheres. The hypothalamus is the source of many of the neurohormones in mammals and is that region of the brain which regulates temperature-control mechanisms in the body.

Hypothermia. A condition in which the body core temperature falls below that considered normal for a homeothermic animal.

Ice nucleation. The induction of ice formation around any small particle that may serve as a nucleous for crystal growth.

Insolation. Incoming solar radiation.

Integument. The outer covering of an insect body.

Isothermal. Being of the same temperature throughout, as from top to bottom of a snowpack or water column.

Isozyme. Multiple forms of the same enzyme, usually having different temperature optima and, thus, varying in their ability to catalyze a reaction at a given temperature.

Intercellular. Between cells (synonomous with Extracellular).

Interscapular. Between the scapulae or shoulder blades.

Intracellular. Within cells.

Latent heat. A quantity of heat tied up or released with phase changes of a substance.

Lewis reaction. Oscillating skin temperatures of the extremities with alternating vasoconstriction and vasodilation induced by cold exposure (also known as the "hunting response").

Lipids. Any of numerous fats or fatlike substances consisting of water-insoluble hydrocarbon molecules that comprise, along with proteins and carbohydrate, the principal structural material of living cells.

Lipoprotein. Proteins combined with lipids (fats).

Littoral zone. The portion of a lake forming the interface between land and deep water, defined by the shoreline at one extent and the limits of submerged, rooted aquatic plants at the other.

Long-wave energy. Radiant energy having a wavelength greater than visible light. In an ecological context usually referring to infrared or heat energy.

Lower critical temperature (LCT). That temperature at which a homeotherm can no longer maintain normal body temperature by passive means and must increase metabolic heat production.

Medulla. Soft marrow-like center or central core (e.g., of bone or hair).

Melanin. A granular pigment varying in color from yellow to black that is present in the feathers and hair of many animals.

Melt metamorphism. A process of change in snowpack structure characterized by melting and refreezing and accompanied by a marked increase in snowpack density.

Metastable. Existing in a nonequilibrium and, therefore, tenuous state for a given set of conditions, as in the case of supercooled water remaining liquid below its normal freezing point.

Microclimate. The climate close to the ground or in the immediate vicinity of an organism (e.g., the nest microclimate).

Microtine. Any rodent in the subfamily *Microtinae,* characteristically having a heavyset body, short legs, and a tail usually shorter than half the head and body length.

Mitochondria. Small rod-shaped or oval organelles in the cell cytoplasm that utilize oxygen and produce most of the ATP, the energy currency of the cell.

Mustelid. Any member of the weasel family, Mustelidae.

Nonshivering thermogenesis. The capacity in homeotherms to increase heat production without muscular activity.

Noradrenaline. A hormone secreted by the adrenal medulla that mediates several thermoregulatory functions, including vasodilation, vasoconstriction, and brown fat metabolism. Also known as norepinephrine.

Overturn. An event or process marked by the reduction of temperature and density differences between the surface and bottom waters of a lake such that complete mixing of the water occurs under the influence of wind-generated currents.

Pelage. The coat (i.e., fur or hair) of a mammal.

Peripheral cooling. A decrease in temperature of external or distal tissues, usually through vasoconstriction, resulting in increased tissue insulation and reduced heat loss.

Permafrost. A thermal condition of earth materials in which the temperature of soil, rock, or water remains below 0°C through the entire year.

Photoperiod. Amount of light received daily. Total length of time between sunrise and sunset.

Phytoplankton. Minute (often microscopic), free-floating plants of marine and aquatic ecosystems.

Piloerection. The erection or fluffing of hairs or feathers resulting in increased insulation (decreased thermal conductivity) through entrapment of air.

Plasma membrane. The semipermeable membrane separating the cytoplasm from the wall of a plant cell.

Poikilotherm. An organism whose body temperature fluctuates with that of its surroundings, except as it may be regulated behaviorally by such means as basking in the sun or through muscular activity. Often referred to as a "cold-blooded" animal.

Polyol. Polyhydric alcohol (an alcohol with three or more hydroxyl (OH) groups).

Radiation. The propagation of energy through space. Also, the energy received by or emitted from a radiating object.

Rete mirabile. "Miraculous net." A dense network of veins and arteries in close contact with each other that facilitates transfer of heat from arterial blood to cold venous blood returning from the extremities, thus reducing heat loss to an animal's surroundings.

Rime ice. Ice that is formed as supercooled water droplets in clouds impact and freeze instantly on exposed objects.

Ring porous. A characteristic ring pattern in the wood (xylem tissue) of some broadleaf tree species in which individual conducting cells vary in size from the conspicuously large diameter and thin-walled cells produced in early spring to the small diameter, thick-walled cells produced in late summer, thus showing a marked line of demarcation between one year's growth and the next (cf. Diffuse porous).

Rumen. The first compartment before the true stomach in certain mammalian digestive systems specialized for the microbial breakdown of cellulose. The rumen serves as a fermentation vat, from which undigested material (the cud) can be regurgitated for additional chewing.

Ruminant. Even-toed, split-hoofed mammals possessing a rumen.

Saturated. In chemistry, having all available valance bonds filled.

Sebaceous gland. A cutaneous gland situated adjacent to hair follicles that secretes oily substances for lubrication of the skin or hair.

Shivering thermogenesis. The capacity in homeotherms to increase heat production by involuntary muscle contractions.

Short-wave energy. Radiant energy having a wavelength within or close to the visible portion of the spectrum. Incoming solar radiation is loosely termed short-wave radiation because most of the sun's energy is emitted within this portion of the spectrum.

Sintering. The fusion of ice grains in contact with each other, tending toward the development of a solid ice mass with concomitant reduction of pore space.

Stomate. Pores on the surface of a leaf formed by a pair of specialized "guard cells" that control the size of the opening to regulate the diffusion of water vapor and carbon dioxide into and out of the leaf.

Subcutaneous. Beneath the skin.

Sublimation. A change in phase from solid directly to vapor.

Subnivean. Beneath the snowpack.

Supercooled. Having a temperature below the normal freezing point of that substance without ice formation occurring.

Sympathetic nerves. System of nerves supplying internal organs and blood vessels and having endings that release adrenaline and noradrenaline.

Temperature gradient. The difference in temperature between two points separated by a given distance. The gradient is said to be "steep" if the difference per unit distance is relatively great.

Temperature inversion. A condition in which air temperature increases with height above the ground, contrary to the normal expectation of a temperature decrease with elevation.

Thermal conductivity. The rate of heat transfer through a material of unit thickness for a given difference in temperature across the material, usually expressed as watts/cm/°K.

Thermal hysteresis. A lag in one of two associated processes or phenomena, as in a given material's having different freezing and melting temperatures.

Thermogenesis. The production of heat in an animal body.

Thermoneutral zone (TNZ). The temperature range within which a homeotherm can maintain its basal metabolic rate through passive regulation of heat gains and losses.

Thermoregulation. Maintenance of body temperature within a preferred range by either physical or physiological means.

Torpor. A short-term condition physiologically similar to hibernation in which metabolic rate and body temperature may be reduced to conserve energy.

Ungulate. A hoofed animal.

Vasoconstriction. Contraction of the circular muscle of arterioles, causing an increase in resistance and decrease in volume of blood flow.

Vasodilation. Opposite of vasoconstriction.

Xylem. The major conducting tissue of vascular plants, comprised mostly of interconnected, nonliving cells through which water moves passively from roots to leaves.

Zooplankton. Minute, drifting marine or aquatic animals having only weak locomotory power.

INDEX